Temperature Measurement in Industry

*An Independent Learning Module
from the
Instrument Society of America*

TEMPERATURE MEASUREMENT IN INDUSTRY

By E. C. Magison

Honeywell Inc.

INSTRUMENT SOCIETY OF AMERICA

INSTRUMENT SOCIETY OF AMERICA
67 Alexander Drive
P.O. Box 12277
Research Triangle Park
North Carolina 27709

Library of Congress Cataloging in Publication Data

Magison, Ernest C.
    Temperature measurement in industry / by E.C. Magison.
        p.   cm.
    Includes bibliographical references.
    ISBN 1-55617-208-7
    1. Temperature measurements.   2. Temperature measurements —
Industrial applications.   3. Engineering instruments.   I. Title.
QC271.M34   1989
681'.2 — dc20   89-26985
                                                          CIP

Editorial development and book design by Monarch International, Inc.,
under the editorial direction of Paul W. Murrill, Ph. D.

TABLE OF CONTENTS

Preface

Unit 1.   **Introduction and Overview**

    1-1.   Course Coverage   3
    1-2.   Purpose   4
    1-3.   Audience and Prerequisites   4
    1-4.   Study Material   5
    1-5.   Organization and Sequence   5
    1-6.   Course Objectives   7
    1-7.   Course Length   7

Unit 2.   **Principles and Fundamentals**

    2-1.   The Objective of Temperature Measurement   11
    2-2.   Direct vs. Indirect Temperature Measurement   12
    2-3.   Temperature Scales   13
    2-4.   Standards and Calibration   15
    2-5.   Types of Temperature Measuring Techniques   16
    2-6.   A Quick Comparison of Measuring Elements   17
    2-7.   Instruments for Indicating, Recording, or Controlling
           Temperature   17

Unit 3.   **Thermoelectric Principles and Thermocouple EMF**

    3-1.   How a Thermocouple Works   21
    3-2.   The Cold Junction   22
    3-3.   The Three Laws of Thermoelectricity   24
    3-4.   Extension and Compensating Extension Wire   25
    3-5.   Thermocouple EMF Calculations   26
    3-6.   Determining Hot Junction Temperature   26
    3-7.   Calibrating an Instrument   29
    3-8.   Checking the Calibration of a Thermocouple   31
    3-9.   Inhomogeneity of Thermocouples   33

Unit 4.   **Thermocouple and Extension Wires**

    4-1.   Thermocouple Types   39
    4-2.   Thermocouple and Extension Wire Insulation   41
    4-3.   Sheathed Thermocouples   45

Unit 5.   **Thermocouples in Use**

    5-1.   The Thermocouple Assembly   51
    5-2.   Protecting Tubes   54
    5-3.   Wells   55
    5-4.   Surface Temperature Measurement   56
    5-5.   Disposable Tip Thermocouples   58
    5-6.   Fabricating Thermocouple Hot Junctions   58
    5-7.   Selecting a Thermocouple for Industrial Temperature
           Measurement   60
    5-8.   Installation Affects Accuracy   61

Unit 6.   **Resistance Thermometers**

     6-1.   General Requirements for Resistance Thermometry   67
     6-2.   Wire-Wound Platinum Elements   67
     6-3.   Thin-Film Platinum Elements   72
     6-4.   Nickel Resistance Thermometers   72
     6-5.   Copper Resistance Elements   73
     6-6.   Characteristics of Resistance Thermometers   74
     6-7.   Measurement Methods   78

Unit 7.   **Filled Systems**

     7-1.   Principle of Operation   85
     7-2.   Classification of Filled Systems   87
     7-3.   Liquid-Filled Systems   87
     7-4.   Vapor Pressure Thermometers   89
     7-5.   Gas-Filled Systems   92
     7-6.   Speed of Response of Filled Systems   93
     7-7.   Application of Filled Systems   93

Unit 8.   **Radiation Thermometers — Theory and Construction**

     8-1.   Why Use Radiation Thermometry?   97
     8-2.   Types of Radiation Thermometers   98
     8-3.   The Theoretical Basis for Radiation Measurements   102
     8-4.   The Response of Radiation Thermometers to Radiation   110
     8-5.   An Overview of Radiation Thermometer Constructions   112
     8-6.   Elements of Radiation Thermometers   117
     8-7.   Field of View   119
     8-8.   Signal Conditioning   122

Unit 9.   **Selection, Application, and Calibration of Radiation Thermometers**

     9-1.   Criteria for Selecting a Radiation Thermometer   127
     9-2.   Application of Radiation Thermometers   129
     9-3.   Radiation Thermometer Accessories   136
     9-4.   Calibration of Radiation Thermometers   136

Unit 10.   **Some Other Methods of Measuring Temperature**

     10-1.   Glass Stem Thermometers   141
     10-2.   Bimetallic Thermometers   141
     10-3.   Pyrometric Indicators   142
     10-4.   Thermistors   144
     10-5.   Semiconductor Sensors   145
     10-6.   Heat Balance Thermometry   146

APPENDIX A: **Suggested Reading and Study Materials**   149
APPENDIX B: **Solutions to All Exercises**   153
Index   163

PREFACE

ISA's Independent Learning Modules

This book is an Independent Learning Module (ILM) as developed and published by the Instrument Society of America. The ILMs are the principal components of a major educational system designed primarily for independent self-study. This comprehensive learning system has been custom designed and created for ISA to more fully educate people in the basic theories and technologies associated with applied instrumentation and control.

The ILM System is divided into several distinct sets of Modules on closely related topics; such a set of individually related Modules is called a Series. The ILM System is composed of:

- the ISA Series of Modules on Control Principles and Techniques;
- the ISA Series of Modules on Fundamental Instrumentation;
- the ISA Series of Modules on Unit Process and Unit Operation Control; and
- the ISA Series of Modules for Professional Development.

The principal components of the Series are the individual ILMs (or Modules) such as this one. They are especially designed for independent self-study; no other text or references are required. The unique format, style, and teaching techniques employed in the ILMs make them a powerful addition to any library.

The published ILMs are as follows:

Fundamentals of Process Control Theory — Paul W. Murrill — 1981

Controlling Multivariable Processes — F. G. Shinskey — 1981

Microprocessors in Industrial Control — Robert J. Bibbero — 1982

Measurement and Control of Liquid Level — Chun H. Cho — 1982

Control Valve Selection and Sizing — Les Driskell — 1983

Fundamentals of Flow Measurement — Joseph P. DeCarlo — 1984

Intrinsic Safety — E. C. Magison — 1984

Digital Control — Theodore J. Williams — 1984

pH Control — Gregory K. McMillan — 1985

FORTRAN Programming — James M. Pruett — 1986

Introduction to Telemetry — O. J. Strock — 1987

Application Concepts in Process Control — Paul W. Murrill — 1988

Controlling Centrifugal Compressors — Ralph Moore — 1989

CIM in the Process Industries — John Bernard — 1989

Continuous Control Techniques for Distributive Control Systems — Gregory K. McMillan — 1989

Temperature Measurement in Industry — E. C. Magison — 1990

Simulating Process Control Loops Using BASIC — Greg Shinskey — 1990

Most of the original ILMs were envisioned to be the more traditional or fundamental subjects in instrumentation and process control. Clearly, with the publications planned over the next few years, the ILM Series will become much more involved in emerging technologies.

Recently, ISA has increased its commitment to the ILM Series and has set for itself a goal of publishing four ILMs each year. Obviously, this growing Series is part of a foundation for any professional library in instrumentation and control. The individual practitioner will find them of value, of course, and they are a necessity in any institutional or corporate library.

There is obvious value in maintaining continuity within your personal set of ILMs; place a standing purchase order with ISA.

Paul W. Murrill
ILM Consulting Editor

Comments about This Volume

This ILM on temperature measurement is designed as a tutorial overview of the most commonly used means of measuring temperature in industrial process measurement and control systems. The objective of this ILM is an understanding of the principles on which the most common sensors operate, the factors that determine the selection and successful use of a sensor for a specific temperature measurement, and an appreciation of the wide range of available sensor hardware.

This textbook, like all of the books in the ILM series, is designed for independent study. It is designed to be useful to salesmen, managers, and supervisors who need only an overview of the subject, as well as to engineers and technicians who will apply the acquired material in their work.

Unit 1:
Introduction
and Overview

UNIT 1

Introduction and Overview

This unit summarizes the content, purpose, objectives, and organization of this Independent Learning Module.

Learning Objectives — When you have completed this unit you should:

A. Understand the organization of the course.

B. Know the course objectives.

C. Know how to proceed through the course.

1-1. Course Coverage

This ILM covers the following subject matter:

- The physical principles underlying the four most common methods used to measure temperature in industrial processes: thermocouples, resistive thermometers, filled systems, and radiation thermometers.

- The varieties of devices available within each of the four important families of measurement techniques.

- The important criteria for selecting a measurement technique to meet the requirements of a process measurement.

- An overview of less common methods of temperature measurement, which may be useful or necessary to meet the needs of a specific application or which may be useful in studying temperature effects in the laboratory or in the plant.

The presentation of this material strives for a balance between principles and practice. On the one hand, this self-study book is not encyclopedic; but on the other, it intends to make clear the wide range of measuring devices that are available and how to choose among them to meet a specific need to measure temperature.

1-2. Purpose

The purpose of the ILM is to help the student who is not now experienced in temperature measurement. Although some courses in measurement and control include some of the material covered in this volume, most individuals in the process measurement and control industry, whether they are employed by manufacturers or users of measuring and control equipment, learn by working with others who are more experienced or by tackling measurement problems themselves. The intent of this book is to speed up the learning process.

1-3. Audience and Prerequisites

This ILM is designed for those who want to work on their own to understand the practical aspects of industrial temperature measurement.

This book will be helpful to anyone needing an understanding of how temperature is measured in industry. For designers of indicating, recording, and control equipment, it will provide a necessary understanding of how such devices interface with temperature measuring devices, and how the functions of these devices relate to measuring systems. In these days of specialization, the material in this book can provide electronics specialists and software engineers an understanding of the measuring equipment with which their products are used.

For engineers and technicians to whom instrumentation is a part of their objective to design, operate, or maintain a production process, this book will contribute to their understanding of the role that temperature measurement can play in a safe, efficient, profitable production facility.

To salesmen, managers, and others in industry whose functions are only tangentially or indirectly concerned with temperature measurement, the material in this ILM will provide an overview that will give them a better understanding of the problems their customers, employees, or colleagues face in resolving problems of temperature measurement.

There are no elaborate prerequisites for use of this ILM. Anyone, no matter how non-technical his education or experience, can benefit from this book. The only mathematical calculations required are simple arithmetic. In Unit 8, some equations are presented whose form may be unfamiliar to some readers. However, these equations are presented only to undergird the presentation. Those who feel uncomfortable with them will find that they are not essential to understanding the important elements of these units.

1-4. Study Material

This textbook, one of ISA's ILM System, is the only study material required for this course. The student who wishes to explore any aspect of temperature measurement in greater detail will find that many vendors of temperature measuring devices and systems provide technical and application notes about the design and use of their products that address specific issues in greater depth.

The instrument technician, the systems engineer, the process engineer, and their supervisors and managers will find it profitable to study other ILMs available from ISA. They present a broad range of material, such as the fundamentals of process control theory, intrinsic safety, control valves, and specific applications of measurement and control.

1-5. Organization and Sequence

This ILM Course is divided into 10 units.

Unit 2 discusses the basic principles and fundamentals of temperature measurement and, in general terms, how process parameters determine the measurement techniques that may be practically and economically justified.

Unit 2 also introduces the student to temperature scales and the system for maintaining calibration standards, and presents a capsule summary of the temperature ranges over which typical thermocouples, resistance thermometers, filled systems, and radiation thermometers are useful.

Unit 3 begins the study of thermocouples with a discussion of how a thermocouple works, cold junction compensation, and the three laws of thermoelectricity. It then describes the use of thermocouple temperature-emf (electromotive force) tables in the calibration of thermocouples and instruments and how to detect error-causing inhomogeneities in thermocouple wire.

Unit 4 looks at the variety of thermocouple types that have been standardized, the kinds of insulation used for thermo-couple and extension wire, and the kinds of hot junction fabri-cation that are relevant to all thermocouple types.

Unit 5 focuses on the use of thermocouples. In this unit, emphasis is on the wide variety of thermocouples designed to meet specific application needs. Included are discussions of protecting tubes and wells and the factors that influence the selection of protecting tube or well material. Unit 5 concludes with a discussion of how the thermocouple installation can affect the accuracy of the measurement.

Unit 6 addresses resistance thermometers, the types available, how the resistance measurement can be interpreted as temper-ature, typical constructions, and common methods of accu-rately measuring their resistance.

Unit 7 teaches the principle of operation of filled thermometer systems, their advantages and disadvantages, the elements of liquid-, gas- or vapor-, and mercury-filled systems, and the characteristics of each that determine its range of application.

Unit 8 begins the study of radiation thermometers with the objective of describing the broad range of radiation thermometer types and the theoretical basis for their design and use. It also presents the characteristics of optical systems and detectors commonly used in constructing radiation thermometer systems as well as field of view specifications and their limitations.

Unit 9 deals with factors a prospective user must consider when selecting a radiation thermometer: how the need of a particular process for speed of response, field of view, and temperature range quickly narrows the choice of thermometer that may be suitable; the common types of applications that may further define the kind of radiation thermometer that can be used; accessories for use with radiation thermometers; and the techniques for calibrating radiation thermometers.

Unit 10 discusses methods of measuring temperature that may be useful in certain applications, especially in laboratory or plant investigations. Many of these techniques can be used only for single determinations of temperature, not for continuous measurement. Others are suitable for continuous measurement but are not widely used because there are no industry standards for their calibration or because they satisfy only special needs.

The method of instruction in this ILM is self-study. You will work at the pace that suits your needs. It is recommended that you omit no unit even though you feel it may not be of importance to you. Rather, at least scan all units and study more intently those of greatest interest.

Each unit is presented in a consistent format. A set of specific learning objectives begins each unit. Read these objectives carefully. The material in the rest of the unit is intended to help you fulfill these objectives.

At the end of each unit are questions and exercises designed to test your understanding of the material. All have answers and solutions in Appendix B. Check your answers and review portions of the unit, if necessary.

1-6. Course Objectives

When you have completed this ILM you should:

A. Understand the internationally recognized system for maintaining temperature scales and how these are reflected in the traceability of the calibration of industrial measuring equipment.

B. Be familiar with the four most commonly used techniques for industrial temperature measurement, the underlying theoretical principles, and common ways in which the principles have been reduced to practical designs.

C. Understand the most common sources of error in the use of these four methods of temperature measurement, how these influence selection among the range of competing designs, and how installation can affect accuracy of measurement.

D. Be familiar with some other ways to measure temperature, which may occasionally be useful, though they are not commonly used for continuous measurement in industry.

1-7. Course Length

The basic philosophy of the ISA system of ILMs is that students learn best if they proceed at their own pace. There is no correct amount of time a student should spend to complete this ILM. The time you take to complete this ILM will depend on your personal needs, your education and experience, and your level of interest.

Though a typical student may complete this ILM in 18 to 24 hours, some will find that their needs for an overview can be met in far fewer hours. Others, who are using this ILM as a first step towards a specialist's expertise in temperature measurement, may spend many more hours in study.

No matter what your motivation for studying this ILM, we recommend an iterative approach. Quickly scan through the entire ILM before beginning more intensive study. This will help you appreciate the place each unit has within the context of the entire ILM. Then, as you study each unit more intensively, it may be helpful to quickly scan each unit before beginning more careful study. In this way, you can better appreciate how each section relates to the entire unit.

Unit 2:
Principles
and Fundamentals

UNIT 2

Principles and Fundamentals

This unit discusses temperature measurement, temperature scales, and standards of calibration that are relevant to all ways of measuring temperature.

Learning Objectives — **When you have completed this unit you should:**

A. Be prepared to select from among temperature measurement techniques the one that is accurate enough and practical for your needs.

B. Know how to convert from one temperature scale to another.

C. Be familiar with the levels of standards calibration for temperature measuring devices.

2-1. What Is the Objective of Temperature Measurement?

In a scientific laboratory, the objective of a measurement may be accuracy regardless of cost and time. This might be the case, for example, when determining the melting point of a new material.

In industry, however, the usual objective is to find a way to measure temperature that yields:

- the accuracy and speed of response required by the application;

- acceptable initial investment; and

- low maintenance cost, which implies long life and stability of calibration.

The following are examples:

- Several types of thermocouples may have similar accuracy and initial cost. The objective is to select the type with greatest stability and longest life. The selection may depend on whether the atmosphere will corrode or poison the thermocouple.

- In glass forming, such as the molding of bottles, viscosity changes require readjustment of the bottle-making machine. Viscosity is closely correlated with temperature, so the accuracy of the temperature measurement is critical. Because glass melts at temperatures too high for most metals to be used and because most materials would contaminate the glass, it is economically justifiable to specify platinum thermocouples in platinum protecting tubes or sophisticated radiation thermometers, an investment that most applications cannot justify.

2-2. Direct vs. Indirect Temperature Measurement

All temperature measurements are indirect, i.e., the measurement is the measurement of volumetric expansion (liquid-filled thermometer), dimensional change (bimetallic thermometers), electromotive force (T/C), resistance, (resistance temperature detector), radiated energy (radiation thermometer), or some other characteristic of a material that varies predictably and reproducibly with temperature.

However, in industrial process measurement and control, the concept of direct and indirect temperature measurement has a different meaning. A direct measurement is a measurement of the temperature of the product itself. An indirect measurement is a measurement of some other temperature from which the product temperature can be inferred.

An example of direct temperature measurement occurs when, as in roasting meat or making candy, it is possible to insert a thermometer directly into the product; insertion or immersion thermometers are often used.

An indirect measurement is used in the baking of bread; oven air temperature is controlled. It is not practical to insert a thermometer into the bread because the quality of the bread would be adversely affected.

Selection of a temperature measurement technique is influenced significantly by the nature of the process: whether it is necessary to measure product temperature and whether it is possible to measure it. It is essential to understand the process and its characteristics before trying to determine which temperature measuring method to use.

For example, it is theoretically possible to devise a measurement scheme to measure the temperature of cookies in a
continuous-bake oven. One could devise a scheme of non-
contact measurement of cookie surface temperature, ignoring
the temperature of the pan. It would be a bad selection because
the process does not depend on surface temperature. The baking
of cookies depends on a time-temperature history of the cookie
dough, and the process is relatively slow. Therefore, measurement of oven temperature is sufficient, simple, reliable, and
economical.

As another example, induction heat treating of a high tech
alloy part may have to be carried out to a very precise end
temperature to achieve the desired physical properties. It may
be essential to measure product temperature directly. A relatively sophisticated radiation pyrometer might be the only way
to make the measurement.

2-3. Temperature Scales

In North America, two temperature scales are in common
industrial use: the Fahrenheit and the Celsius scales.

The Fahrenheit scale was invented by Daniel Gabriel Fahrenheit and was published in 1724. The ice point was designated
32°F, and the steam point of water was 212°F. It is still used
extensively in North America and other English speaking countries, though even in these countries many industries are
slowly converting to the Celsius scale.

The Celsius scale was originated in 1742 by Anders Celsius of
Uppsala, Sweden. He designated the ice point as 100°C and the
steam point as 0°C. Several years later the designations were
reversed. Today the ice point is 0°C and the steam point is
100°C.

The Kelvin Scale is used primarily in scientific work, less frequently in industrial measurement. The ice point is defined as
273.16 K. 1 K = 1°C.

The Rankine scale is the analog of the Kelvin Scale, using Fahrenheit degrees. The ice point is 491.69° Rankine. 1°R = 1°F.
The low end of both the Kelvin and Rankine scales is zero
degrees. Thus, both the Kelvin and Rankine scales are called
absolute scales. In theory, at 0 K or 0°R, all molecular motion
ceases.

The following will illustrate conversion from one scale to the other.

Degrees Celsius and kelvins:

$$°C = K - 273.16$$
$$K = °C + 273.16$$

Examples:

$$25°C = 25°C + 273.16 = 298.16 \text{ K}$$
$$100 \text{ K} = 100 - 273.16 = -173.16°C$$

Degrees Fahrenheit and degrees Rankine:

$$32°F = 491.60° \text{ Rankine, therefore,}$$
$$°F = °R - 459.69$$
$$°R = °F + 459.69$$

Degrees Celsius and degrees Fahrenheit:

$$°C = \frac{5}{9} (°F - 32)$$

$$°F = °C \frac{9}{5} + 32$$

Examples:

$$25°C = 25 \times \frac{9}{5} + 32 = 45 + 32 = 77°F$$

$$100°F = \frac{5}{9} (100 - 32) = \frac{5}{9} \times 68 = 37.8°C$$

Degrees Rankine and kelvins:

Because both scales start at zero and the ice point is 491.69°R and 273.16 K,

$$°R = \frac{491.69 \text{ K}}{273.16} = 1.8 \text{ K}$$

$$K = \frac{°R}{1.8}$$

Alternatively, convert K to °C, then convert °C to °F and add 459.69 to obtain °R.

Since 1887 there have been a series of international agreements about how to define the Celsius and Kelvin scales. In 1927 six fixed points were established, together with specifications for the interpolation instruments, the platinum resistance thermometer, and the platinum 10% rhodium-platinum thermocouple. The scale was changed slightly in 1948 and again in 1968.

Eleven fixed points extend from the triple point of hydrogen, –259.34°C, to the freezing point of gold, 1064.43°C. The triple point of a material is the temperature at which the solid, liquid, and vapor phases are in equilibrium.

National standards of all countries are referenced to the International Practical Temperature Scale of 1968 (IPTS68). Standard thermocouple and RTD calibration curves are also referred to IPTS68.

2-4. Standards and Calibration

National standards laboratories maintain primary standards for interpolating between the fixed points. The standard platinum resistance thermometer, Pt-Pt 10% rhodium thermocouple and the standard optical pyrometer are calibrated at two or more of the fixed points. They are then used as standards for intermediate or higher temperatures. These national standards devices are primary reference standards. Devices for frequent use are calibrated against the primary standards and used for routine work so that the effects of damage or destruction are less serious.

There are 5 recognized levels of standards calibration:

- Level 5—International standards coordination: cross verification of primary standards against fixed points and standards of other laboratories.

- Level 4—Primary standards calibration system:

    (1) Usually consists of metrology grade platinum resistance thermometers, a high quality precision resistance bridge and null detector, a calibrated standard resistor to check

the bridge; a set of primary standard freezing point cells and precision-stirred liquid calibration baths.

(2) A regular program of verification to insure accuracy of the resistance thermometers and the reference resistor.

(3) Continuous records of verification of standards and their calibrations.

- Level 3—Secondary reference standards are usually maintained by manufacturers. Depending on the temperature range of interest, they may be precision mercury-in-glass thermometers, platinum resistance thermometers, platinum-10% rhodium-platinum thermocouples, or optical pyrometers calibrated against primary standards.

- Level 2—Industrial grade measuring devices, accuracy not necessarily built in, but stable enough so that occasional checks and adjustments based on secondary reference standards can maintain the required accuracy.

- Level 1—Devices where the effects of error are not severe, e.g., a home thermostat where the setting can be changed easily if the reading is off a few degrees; or in an oven where the baking time can be adjusted if the temperature is incorrect, or the oven temperature can be adjusted based on observation of the product.

2-5. Types of Temperature Measuring Techniques

The remainder of this ILM will consider in some detail the following types of temperature measuring elements, which are the most commonly used in industrial applications:

- Thermocouples
- Resistance Thermometers
- Filled Systems
- Radiation Pyrometers

Unit 10 will describe briefly other methods that are not frequently used for continuous measurements or which are limited to a few specific applications.

2-6. A Quick Comparison of Measuring Elements

The range of temperatures over which the four types of measuring elements are commonly used is shown in Fig. 2-1. Sensors may sometimes be used outside those ranges, but life and stability of calibration may be adversely affected. The ranges shown for radiation pyrometers include a wide variety of models and designs. No single design functions well over the entire range.

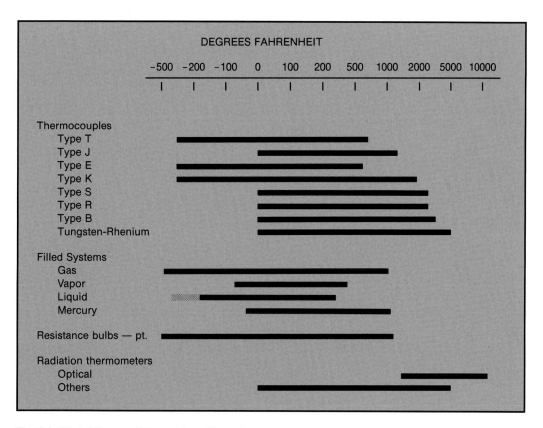

Fig. 2-1. Useful Range of Temperature Elements

2-7. Instruments for Indicating, Recording, or Controlling Temperature

This ILM does not discuss in detail the instruments used to measure, indicate, or record electromotive force (emf) or resistance output of temperature sensors.

The current state of the electronic art provides a very broad range of choice of instruments, analog or digital, with accuracy of 0.1 to 1.0% of span. Control options are available, ranging from simple on-off control to sophisticated algorithms, and with outputs suitable for driving all sorts of loads. Therefore, once a temperature measuring method has been selected, it is relatively easy to select the indicator, recorder, or controller with accuracy, features, and price suitable for the specific application.

Exercises

2-1.   *Make the following conversions:*

(a) *953°C* = _____ *°F*
(b) *1000°F* = _____ *K*
(c) *1800 K* = _____ *°F*
(d) *1800 K* = _____ *°R*

2-2.   *Why is a well stirred mixture of ice and water used as a reference temperature in industrial calibration practice?*

2-3.   *What temperature measuring devices might be considered to measure a temperature of:*

(a) *–100°F*
(b) *2000°F*

Unit 3:
Thermoelectric Principles and Thermocouple EMF

UNIT 3

Thermoelectric Principals and Thermocouple EMF

This unit deals with the basic principles and laws of thermo-electricity, the configuration of the usual thermocouple circuit, and the use of the temperature-emf table for a thermocouple.

Learning Objectives — When you have completed this unit you should:

A. Know how to determine hot and cold junction temperatures from the thermocouple electromotive force (emf) reading.

B. Know how to calculate emf readings for checking instrument calibration.

C. Know how to calibrate an instrument with and without an ice bath.

D. Understand the use of extension and compensating wires.

3-1. How a Thermocouple Works

A thermocouple is an assembly of two wires of unlike metals joined at one end, the "hot" or measuring junction. At the other end, the "cold" junction, (see Fig. 3-1), the open circuit voltage is measured. This voltage, the Seebeck voltage, depends on the difference in temperature between the hot junction and the cold junction and the Seebeck coefficient of the two metals. The current, I, is the Seebeck current, which will flow if terminals a and b at the cold junction are shorted together.

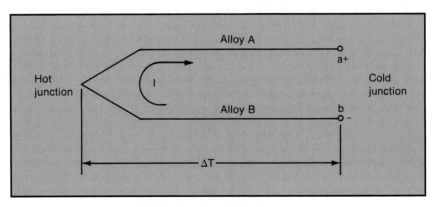

Fig. 3-1. Thermocouple Circuit — Seebeck Effect

Another characteristic of thermocouples, the Peltier effect, is not used in temperature measurement but is the basis for thermoelectric heating and cooling. A current flowing across the junction of two dissimilar metals causes heat to be absorbed (if the current is in the same direction as the Seebeck current) and liberated (if the current flows opposite to the Seebeck current).

The Thompson effect, Fig. 3-2, occurs in a single conductor. If the conductor is heated at a point to a temperature T_1, two points, P_1 and P_2, on either side will be at a lower temperature, T_2. If current I flows in the wire as shown, electrons absorb energy at P_2, as the current flows opposite to the temperature gradient. This same energy is released at P_1, as the current flows in the same direction as the temperature gradient. Because the gain and loss are equal, there is no net effect along the wire. The practical significance of this Thomson effect is that application of heat to a single homogeneous wire does not generate a net thermoelectric voltage. This fact is the basis for connecting measuring devices to thermocouples with copper wires, because they themselves do not add voltage to the circuit.

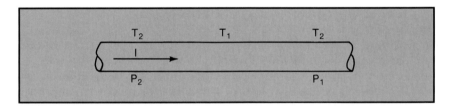

Fig. 3-2. The Thompson Effect

3-2. The Cold Junction

The Seebeck emf depends on the characteristics of the thermo-couple metals and the temperature difference between the measuring and cold junctions. A measurement of emf can be interpreted as the temperature of the hot junction only if the cold junction temperature is known or if any changes in cold junction temperature are compensated for in the measuring circuitry.

In the laboratory a common technique is to use an ice bath to maintain the cold junction at 32°F (0°C). Copper wires are joined to the thermocouple. The junctions are insulated and placed in a well stirred bath of ice and water (see Fig. 3-3).

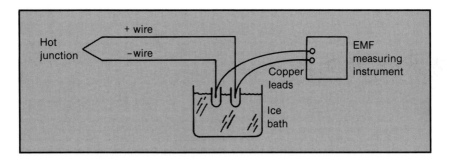

Fig. 3-3. Use of Ice Bath To Maintain Constant Cold Junction Temperature

Because the copper is homogeneous, a temperature gradient along the copper wire generates no emf. Because both conductors are the same material, the difference between ice bath temperature and the temperature at the terminals of the measuring instrument also does not produce an emf.

Rather than use an ice bath, a reference "oven" may be used to maintain the cold junction at a constant known temperature. The copper wire connections may be made in the same way. The oven may be cooled below ambient by a thermo-electric cooler, or it may be heated and controlled at a constant temperature above ambient temperature.

In measuring and control instrumentation, maintaining a constant cold junction temperature is inconvenient. Some measuring instruments use a cold junction compensating resistor (R_T) to automatically compensate for cold junction temperature changes (see Fig. 3-4). The cold junction resistor is at cold junction temperature and is usually sized so that the emf from the voltage divider is zero at a reference ambient of 75°F (24°C). If the cold junction temperature increases, thermocouple emf decreases, but the cold junction resistor increases resistance, adding an emf in series with the thermocouple that is equal to the decrease in thermocouple emf. The measuring

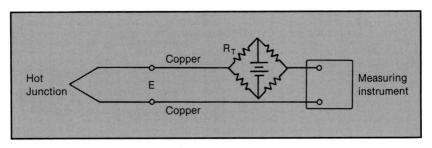

Fig. 3-4. Automatic Cold Junction Compensation

instrument consequently sees a constant emf, regardless of changing ambient temperature.

In digital instruments, compensation for changes in cold junction temperature may be implemented differently. In the scheme illustrated in Fig. 3-4, the incremental emf caused by cold junction temperature changes is added or subtracted directly to or from the thermocouple emf. In digital instruments the cold junction compensating resistor may be supplied with a small constant current. The variation of voltage across the resistor caused by temperature changes is converted to digital form, which is then added to or subtracted from the digital value that represents the thermocouple emf at the input terminals.

3-3. The Three Laws of Thermoelectricity

Experience has shown that the behavior of thermocouple circuits can be summarized by three laws.

1. Law of the Homogeneous Circuit. In a circuit of a single homogeneous metal, however varied the cross section, an electric current cannot be sustained by the application of heat alone.

2. Law of Intermediate Metals (Fig. 3-5). If, in a circuit of dissimilar metals, homogeneous wire of a third metal is introduced between points X and Y, no additional emf will be generated if points X and Y are at the same temperature.

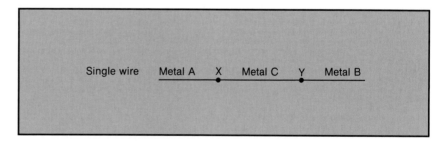

Fig. 3-5. Law of Intermediate Metals

3. Law of Intermediate Temperature (Fig. 3-6). In a thermocouple circuit, the emf generated by the temperature difference $T_1 - T_3$ is equal to the emf generated by the temperature difference $T_1 - T_2$ plus the emf generated by the temperature difference $T_2 - T_3$.

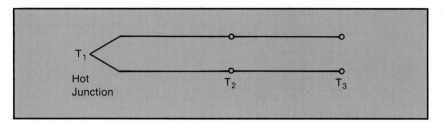

Fig. 3-6. Law of Intermediate Temperature

3-4. Extension and Compensating Wire

A typical industrial thermocouple circuit is shown in Fig. 3-7. Thermocouple wire is used from the hot junction to the terminals of the thermocouple assembly. Extension wire or compensating wire is run from the terminals of the thermocouple assembly to the measuring instrument. Extension wire or compensating wire is in general less expensive than thermocouple wire because:

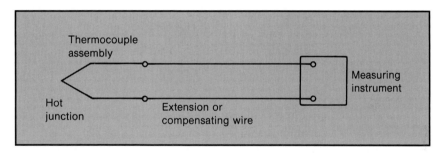

Fig. 3-7. Typical Industrial Thermocouple Circuit

- it doesn't need high temperature insulation (the thermocouple terminals are usually at 400°F (204°C) or below so the extension wire insulation need be rated for only 400°F);

- accuracy is required over a relatively narrow temperature range (though thermocouple wire accuracy is controlled to 1600°F (871°C), for example, extension wire accuracy need be controlled only to 400°F (204°C); and

- wire size can be chosen to suit installation requirements (the thermocouple may use 8-gauge wire for long life and stability, and the extension leads may need to be only 18-gauge to be strong enough for easy reliable installation).

Extension wire is constructed of the same materials as the thermocouple and, within its rated temperature range, has the same accuracy. Extension wires are used with base metal thermocouples (see Tables 4-1 and 4-2).

Compensating wire is not made of the same materials as the corresponding thermocouple wire. It is used with noble metal (platinum) and refractory metal (tungsten) thermocouples to reduce cost and because the physical properties of the wire are superior to those of tungsten alloys. Its accuracy over the usable temperature range differs from that of the thermocouple wire (see Table 4-3).

3-5. Thermocouple EMF Calculations

Listed in Table 3-1 are the partial temperature-emf characteristics of a Type K, Chromel-Alumel™ thermocouple. Note that at 32°F (0°C) the emf is zero. This means that the table is based on a cold or reference junction temperature of 32°F. All standard temperature-emf tables are based on 32°F (0°C) because of the relative ease of maintaining this temperature with a mixture of crushed ice and water.

Three kinds of calculations using the temperature-emf table are necessary when using a thermocouple measuring system.

- Determine the temperature of the hot junction from a measurement of emf at the cold junction.

- Find the amount of emf to apply to a measuring instrument to check its calibration.

- Check the calibration of a thermocouple.

3-6. Determining Hot Junction Temperature

The following procedure must be performed in order to determine the hot junction temperature.

- Measure the emf at the cold junction.

- Measure the temperature at the cold junction, T_{cj}.

Chromel and Alumel are registered trademarks of Hoskins Manufacturing Co.

°F	0	1	2	3	4	5	6	7	8	9	10	°F
0	-0.692	-0.671	-0.650	-0.628	-0.607	-0.585	-0.564	-0.543	-0.521	-0.500	-0.478	0
10	-0.478	-0.457	-0.435	-0.413	-0.392	-0.370	-0.349	-0.327	-0.305	-0.284	-0.262	10
20	-0.262	-0.240	-0.218	-0.197	-0.175	-0.153	-0.131	-0.109	-0.088	-0.066	-0.044	20
30	-0.044	-0.022	0.000	0.022	0.044	0.066	0.088	0.110	0.132	0.154	0.176	30
40	0.176	0.198	0.220	0.242	0.264	0.286	0.308	0.331	0.353	0.375	0.397	40
50	0.397	0.419	0.441	0.464	0.486	0.508	0.530	0.553	0.575	0.597	0.619	50
60	0.619	0.642	0.664	0.686	0.709	0.731	0.753	0.776	0.798	0.821	0.843	60
70	0.843	0.865	0.888	0.910	0.933	0.955	0.978	1.000	1.023	1.045	1.068	70
80	1.068	1.090	1.113	1.135	1.158	1.181	1.203	1.226	1.248	1.271	1.294	80
90	1.294	1.316	1.339	1.362	1.384	1.407	1.430	1.452	1.475	1.498	1.520	90
100	1.520	1.543	1.566	1.589	1.611	1.634	1.657	1.680	1.703	1.725	1.748	100
110	1.748	1.771	1.794	1.817	1.839	1.862	1.885	1.908	1.931	1.954	1.977	110
120	1.977	2.000	2.022	2.045	2.068	2.091	2.114	2.137	2.160	2.183	2.206	120
130	2.206	2.229	2.252	2.275	2.298	2.321	2.344	2.367	2.390	2.413	2.436	130
140	2.436	2.459	2.482	2.505	2.528	2.551	2.574	2.597	2.620	2.643	2.666	140
150	2.666	2.689	2.712	2.735	2.758	2.781	2.804	2.827	2.850	2.873	2.896	150
160	2.896	2.920	2.943	2.966	2.989	3.012	3.035	3.058	3.081	3.104	3.127	160
170	3.127	3.150	3.173	3.196	3.220	3.243	3.266	3.289	3.312	3.335	3.358	170
180	3.358	3.381	3.404	3.427	3.450	3.473	3.496	3.519	3.543	3.566	3.589	180
190	3.589	3.612	3.635	3.658	3.681	3.704	3.727	3.750	3.773	3.796	3.819	190
200	3.819	3.842	3.865	3.888	3.911	3.934	3.957	3.980	4.003	4.026	4.049	200
210	4.049	4.072	4.095	4.118	4.141	4.164	4.187	4.210	4.233	4.256	4.279	210
220	4.279	4.302	4.325	4.348	4.371	4.394	4.417	4.439	4.462	4.485	4.508	220
230	4.508	4.531	4.554	4.577	4.600	4.622	4.645	4.668	4.691	4.714	4.737	230
240	4.737	4.759	4.782	4.805	4.828	4.851	4.873	4.896	4.919	4.942	4.964	240
250	4.964	4.987	5.010	5.033	5.055	5.078	5.101	5.124	5.146	5.169	5.192	250
260	5.192	5.214	5.237	5.260	5.282	5.305	5.327	5.350	5.373	5.395	5.418	260
270	5.418	5.440	5.463	5.486	5.508	5.531	5.553	5.576	5.598	5.621	5.643	270
280	5.643	5.666	5.688	5.711	5.733	5.756	5.778	5.801	5.823	5.846	5.868	280
290	5.868	5.891	5.913	5.936	5.958	5.980	6.003	6.025	6.048	6.070	6.092	290
300	6.092	6.115	6.137	6.160	6.182	6.204	6.227	6.249	6.271	6.294	6.316	300
310	6.316	6.338	6.361	6.383	6.405	6.428	6.450	6.472	6.494	6.517	6.539	310
320	6.539	6.561	6.583	6.606	6.628	6.650	6.672	6.695	6.717	6.739	6.761	320
330	6.761	6.784	6.806	6.828	6.850	6.873	6.895	6.917	6.939	6.961	6.984	330
340	6.984	7.006	7.028	7.050	7.072	7.094	7.117	7.139	7.161	7.183	7.205	340
350	7.205	7.228	7.250	7.272	7.294	7.316	7.338	7.361	7.383	7.405	7.427	350
360	7.427	7.449	7.471	7.494	7.516	7.538	7.560	7.582	7.604	7.627	7.649	360
370	7.649	7.671	7.693	7.715	7.737	7.760	7.782	7.804	7.826	7.848	7.870	370
380	7.870	7.893	7.915	7.937	7.959	7.981	8.003	8.026	8.048	8.070	8.092	380
390	8.092	8.114	8.137	8.159	8.181	8.203	8.225	8.248	8.270	8.292	8.314	390
400	8.314	8.336	8.359	8.381	8.403	8.425	8.448	8.470	8.492	8.514	8.537	400
410	8.537	8.559	8.581	8.603	8.626	8.648	8.670	8.692	8.715	8.737	8.759	410
420	8.759	8.782	8.804	8.826	8.849	8.871	8.893	8.916	8.938	8.960	8.983	420
430	8.983	9.005	9.027	9.050	9.072	9.094	9.117	9.139	9.161	9.184	9.206	430
440	9.206	9.229	9.251	9.273	9.296	9.318	9.341	9.363	9.385	9.408	9.430	440
450	9.430	9.453	9.475	9.498	9.520	9.543	9.565	9.588	9.610	9.633	9.655	450
460	9.655	9.678	9.700	9.723	9.745	9.768	9.790	9.813	9.835	9.858	9.880	460
470	9.880	9.903	9.926	9.948	9.971	9.993	10.016	10.038	10.061	10.084	10.106	470
480	10.106	10.129	10.151	10.174	10.197	10.219	10.242	10.265	10.287	10.310	10.333	480
490	10.333	10.355	10.378	10.401	10.423	10.446	10.469	10.491	10.514	10.537	10.560	490
500	10.560	10.582	10.605	10.628	10.650	10.673	10.696	10.719	10.741	10.764	10.787	500
510	10.787	10.810	10.833	10.855	10.878	10.901	10.924	10.947	10.969	10.992	11.015	510
520	11.015	11.038	11.061	11.083	11.106	11.129	11.152	11.175	11.198	11.221	11.243	520
530	11.243	11.266	11.289	11.312	11.335	11.358	11.381	11.404	11.426	11.449	11.472	530
540	11.472	11.495	11.518	11.541	11.564	11.587	11.610	11.633	11.656	11.679	11.702	540
°F	0	1	2	3	0	5	6	7	8	9	10	°F

Converted from °C (IPTS 1968).

TABLE 3-1. Partial Type K Thermocouple Temperature — EMF Table

- Determine the emf at T_{cj} from the table. This is the emf the thermocouple would read if its hot junction were in an ice bath.

- Add this emf to the emf measured in the first step. This emf would be developed by the thermocouple if the cold junction were in an ice bath and the hot junction is at the unknown temperature.

- Find this emf in the table and read the corresponding temperature. This is the hot junction temperature.

Some examples will illustrate this procedure.

(1) No interpolation required.

- The emf measured at the cold junction is 10.333 mv.

- The temperature of the cold junction is 74°F.

- From Table 3-1, the emf of the cold junction is 0.933 mv.

- The total emf referred to 32°F is 10.333 + 0.933 = 11.266 mv.

- From Table 3-1, this is equivalent to a hot junction temperature of 531°F.

(2) Interpolation required.

- The emf measured at the cold junction is 9.117 mv.

- The temperature at the cold junction is 69.4°F.

- Table 3-1 has values of emf only for integer values of temperature, so it is necessary to use linear interpolation. Linear interpolation assumes that the emf characteristic is linear between adjacent temperature values, so the emf for 69.5°F, for example, is halfway between the emf for 69°F and the emf for 70°F.

- For 69.4°F the emf is $\dfrac{69.4 - 69}{70 - 69} (mv_{70} - mv_{69}) + mv_{69}$.

This equals $\dfrac{0.4}{1} \times (0.843 - 0.821) + 0.821 = (0.4 \times 0.022) + 0.821$
$$= 0.8298 \text{ mv, rounded}$$
$$\text{off to } 0.830 \text{ mv.}$$

- The total emf referred to 32°F is 9.117 + 0.830 = 9.947 mv. This value is not found in Table 3-1.

- For T_{HJ} of 473°F the emf is 9.948 mv. For T_{HJ} of 472°F the emf is 9.926 mv.

$$T_{HJ} = 472 + \frac{9.947 - 9.926}{9.948 - 9.926} \times (473 - 472)$$

$$= 472 + \frac{0.021}{0.022} \times 1$$

$$= 472 + 0.95$$

$$= 472.95°F$$

3-7. Calibrating an Instrument

One usually wants to check at several points within the cali-brated range as well as at the lower range and upper range limits of an instrument.

Without an ice bath, the procedure is as follows:

- Measure the temperature at the cold junction.

- Read the emf corresponding to the cold junction tempera-ture from the temperature-emf table.

- Read the emf corresponding to the lower range limit from the temperature-emf table.

- Subtract the cold junction temperature emf from the lower range limit emf.

- Apply the resulting emf to the input terminals and read the indicated temperature.

- Calculate the error.

- Repeat for other points on scale.

Example 3-1. An instrument is calibrated 0–500°F Type K. It is desired to check its calibration at 0°, 100°, 200°, 300°, 400°, and 500°F. The cold junction temperature is measured at 78°F. From Table 3-1 one finds mv_{CJ} = 1.023 mv.

The values of emf at the various temperatures corresponding to a cold junction temperature of 32°F are:

0°F	−0.692 mv
100°F	1.520 mv
200°F	3.819 mv
300°F	6.092 mv
400°F	8.314 mv
500°F	10.560 mv

Subtracting 1.023 mv from each of these values, we find that the input voltages should be:

0°F	−1.715 mv
100°F	0.497 mv
200°F	2.796 mv
300°F	5.069 mv
400°F	7.291 mv
500°F	9.537 mv

Assume that the instrument indication (or record) for each of these inputs is 1°F, 102°F, 203°F, 304°F, 405°F, 506°F, respectively. The errors are, therefore:

0°F	+1°F	0.2% span (span = 500°F)
100°F	+2°F	0.4% span
200°F	+3°F	0.6% span
300°F	+4°F	0.8% span
400°F	+5°F	1.0% span
500°F	+6°F	1.2% span

With an ice bath (Fig. 3-9), the procedure is as follows:

- Use an ice bath with copper leads from the source of emf to the ice bath and extension leads from the ice bath to the instrument under test.

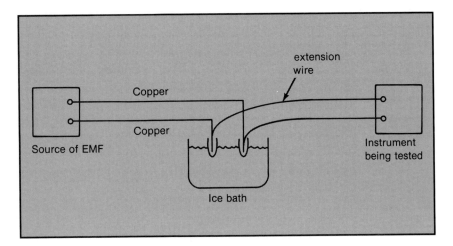

Fig. 3-9. Calibration of an Instrument Using Ice Bath

- Read the emf from the table corresponding to the desired calibration point.

- Apply this emf to the copper leads. The extension leads will subtract the emf due to the difference between the instrument cold junction temperature and 32°F.

Unless the accuracy of the extension leads is known, this method may not be more accurate than the previous method.

Example 3-2. The instrument of the previous example is to be calibrated using an ice bath and extension leads. Because of the automatic subtraction by the extension leads of an emf proportional to the cold junction temperature, it is necessary only to read from Table 3-1 the emf's for each point on scale to be checked.

0	−0.692 mv
100	1.520 mv
200	3.819 mv
300	6.092 mv
400	8.314 mv
500	10.560 mv

3-8. Checking the Calibration of a Thermocouple

The best way to check a thermocouple's calibration is to measure its emf differentially against the emf of a thermocouple with known accuracy (see Fig. 3-10).

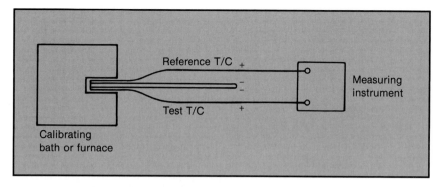

Fig. 3-10. Calibrating a Thermocouple

It is easier to measure small differences of emf, perhaps a few microvolts, than to measure a large emf with high accuracy.

A small error in the calibration temperature does not affect the accuracy of the calibration significantly. For example, if two Type K thermocouples differ by 3°F at 1000°F, the error will be essentially the same at 995° or 1005°F.

The method is as follows:

- Use a reference thermocouple of the same type, which has been calibrated carefully against a standard of known accuracy.

- Twist the hot junctions of the reference and test thermocouples together to ensure that they will be at the same temperature, or place them in the same small tube or well.

- Ensure that the cold junctions are at the same temperature, for example, by inserting each twisted pair in a tube in an ice bath or in a block of metal of high mass.

- Set the calibration bath or furnace to the desired temperature.

- Measure the differential emf and record.

- Compute the absolute error, taking into account the known inaccuracy of the reference thermocouple.

Example 3-3. A factory standard Type K thermocouple is used to check the calibration of a Type K thermocouple that has been in use and is suspected of having drifted. The calibration is to be checked at 500°F. The factory standard has been calibrated and has an error of + 1°F = + 0.022 mv at 500°F. The two hot junctions are twisted together and the thermocouples are differentially connected as in Fig. 3-10. The measuring instrument reads + 0.056 mv. What is the error of the thermocouple under test?

The + 0.056 mv reading means that the factory standard thermocouple has a higher emf than the test thermocouple. But the factory standard thermocouple has an error of + 0.022 mv relative to the nominal calibration (Table 3-1). The test thermocouple, therefore, is below the nominal calibration curve by 0.056 – 0.022 mv or 0.034 mv. At 500°F it would read 0.034 mv low. The nominal emf at 500°F is 10.560 mv. It would read 10.526 mv. Reference to Table 3-1 shows that this is equivalent to 498.5°.

Another way to compute the error in °F is to note that at 500°F a 1°F change in temperature causes a change of 0.022 to 0.023 mv.

500°F 10.560
$\qquad$ = 0.022
501°F 10.582
$\qquad$ = 0.023
499°F 10.537

Therefore, error = $\dfrac{-0.034}{0.023}$ = –1.5°F

Another way to check the calibration of a thermocouple is to use a reference standard platinum thermocouple to determine the temperature of the furnace by measuring its emf. Then measure the emf of the test thermocouple and determine the error based on the known temperature. This method requires a more accurate measurement technique because the platinum thermocouple emf changes only 5–6 microvolts per °F.

3-9. Inhomogeneity of Thermocouples

The law of the homogeneous circuit is valid only if the material composition is same at every point along the wire. Though the wire may be homogeneous when it is purchased from the manufacturer, handling and fabrication can induce strain, which changes the Seebeck coefficient locally. Inhomogeneity may also be induced by local corrosion, which changes the composition of the wire.

Inhomogeneity can produce errors. If there is a significant temperature gradient along the wire so that there is a temperature difference between an inhomogeneous zone in a wire and the adjacent zone, an emf will be generated. If a thermocouple is removed from service for calibration and the gradient in the calibrating furnace is not the same as that in the process, the emf output may be different even though the hot junction is at the same temperature in both the process and the calibrating furnace.

The following procedure should be used in testing for inhomogeneity (Fig. 3-11).

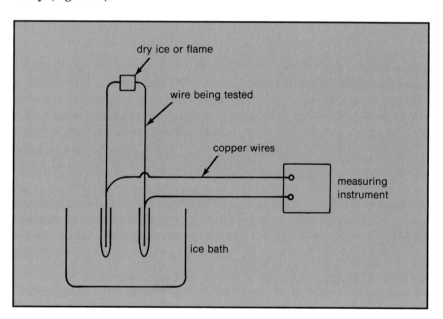

Fig. 3-11. Checking Homogeneity of Thermocouple Wire

- Place the ends of the wire being tested in an ice bath and connect copper wires to the measuring instrument.

- Draw a piece of dry ice or a small flame along the wire to generate a local temperature gradient. If the wire is inhomogeneous, the measuring instrument will indicate an emf.

- If inhomogeneity is caused by the strain of mechanical working, it can be reduced by annealing.

- There is no practical way to compensate for the effects of inhomogeneity. If the emf cannot be reduced by annealing, then the thermocouple must be replaced.

Exercises

3-1. Which thermoelectric effects are most relevant to the user of thermocouples for temperature measurement?

3-2. Why is it necessary to maintain the cold junction at a known constant temperature or else compensate in the measuring system for changes in cold junction temperature if the measuring system is calibrated in temperature units?

3-3. What is the law of the homogeneous circuit?

3-4. What is the law of intermediate metals, and why is it important?

3-5. What is the law of intermediate temperature?

3-6. Why are extension and compensating wire used?

3-7. The emf of a Type K thermocouple is found to be 8.054 mv when the cold junction temperature is 67°F. What is the hot junction temperature?

3-8. If an ice bath and extension wire are used to calibrate an instrument at 400°F (Type K thermocouple), what emf must be supplied by the calibrating source?

3-9. If an ice bath and extension wire are not used to calibrate the instrument of Exercise 3-8 at 400°F, what emf output should be set on the source? $T_{cj} = 69°F$?

3-10. A Type K thermocouple is removed from service and calibrated differently against a reference Type K thermocouple at a temperature of 500°F. The connections are made according to Fig. 3-10. The reference thermocouple is known to be 1.3°F low at 500°F. The measuring instrument reads −0.010 mv. What is the accuracy of the working thermocouple that is being checked?

3-11. If the thermocouple under test of Exercise 3-10 shows a large error, perhaps 10–20°F, the cause may be inhomogeneity of the wires due to corrosion. How can this possibility be verified?

Unit 4:
Thermocouple and Extension Wires

UNIT 4

Thermocouple and Extension Wires

This unit reviews the available kinds of thermocouple wires with standardized calibration curves, their initial accuracy limits, and the limitations on accurate and reliable use.

The types of thermocouple insulation materials in common use are also presented, as are the characteristics and construction of extension and compensating wire.

The objective of this unit is to provide an overview of the characteristics of thermocouple and extension wires as background for Unit 5, which considers how one selects a thermocouple type and construction.

Learning Objective — **When you have completed this unit you should:**

A. **Have a working knowledge of the various types of thermocouple and extension wires and the materials with which they can be insulated.**

4-1. Thermocouple Types

Table 4-1 shows the usable temperature ranges and limits of error of the most common thermocouple types.

Thermocouple Type	Temperature Range	Limits of Error	
		Standard*	Premium
T	0–350° C, 0–660° F	±1° C or ±0.75%	±.5° C or 0.4%
	–200–0° C, –300° –32° F	±1° C or ±1.5%	—
J	0–750° C, 0–1400° F	±2.2° C or ±0.75%	±1.1° C or ±0.4%
E	0–900° C, 0–1650° F	±1.7° C or ±0.5%	±1° C or ±0.4%
	–200° –0° C, –300° –32° F	±1.7° C or ±1%	—
K	0–1250° C, 0–2300° F	±2.2° C or ±.75%	±1.1% or ±0.4%
	–200° –0° C, –300° –32° F	±2.2° C or ±2%	—
R,S	0–1450° C, 32–2700° F	±1.5° C or ±.25%	±0.6° C or ±0.1%
B	800–1700° C, 1500–3100° F	±0.5%	—
Tungsten-Rhenium 24 ga.**	0–2930° C, 32–4200° F	±13.3° C or 1%	—

* Whichever is greater
** Smaller gauges have lower temperature limits

TABLE 4-1. Operating Range and Limits of Error of Common Thermocouple Types

- Type J, Iron-Constantan

Type J thermocouples are not used below 0°C because condensation may cause rusting of the iron leg. They are best used in reducing atmospheres. The highest operating temperature depends on wire size. The maximum temperature indicated in Table 4-1 applies only to large wire. Type J thermocouples can be used in dry reducing atmospheres to 400°C.

- Type T, Copper-Constantan

Type T thermocouples may be used in slightly oxidizing or reducing atmospheres and are frequently used for low temperature work. Both wires can easily be annealed to reduce the errors caused by inhomogeneity in large temperature gradients. Type T resists moisture well.

- Type K, Chromel-Alumel™

Use Type K thermocouples only in an oxidizing atmosphere. A reducing atmosphere, as found inside a well, may cause green rot, which shifts calibration. Green rot results from migration of chromium. Sulfur compounds will also attack Type K wire. Although the Type K thermocouple was originally Chromel-Alumel™, other alloys are now available that fit the standard temperature-emf curve. Some give more stable calibration at higher temperature and greater resistance to green rot. Because these alloys are of palladium, platinum, and gold, they are more costly.

- Type E, Chromel-Constantan

Type E thermocouples have the highest emf output of the standard thermocouple types. Type E can be used at sub-zero temperatures. It should not be used in strongly reducing atmospheres and must be protected from sulfur compounds.

- Types R, S, B, Platinum-Rhodium (Type S is the International Transfer Standard for the IPTS 68.)

Type	Positive Wire	Negative Wire
S	90% platinum, 10% rhodium	Platinum
R	87% platinum, 13% rhodium	Platinum
B	70% platinum, 30% rhodium	94% platinum 6% rhodium

These types of thermocouples have excellent resistance to oxidizing atmospheres, but poor resistance to reducing atmospheres. Above 1000°C platinum is contaminated by hydrogen, carbon, and metallic vapors. At high temperatures, ceramic insulation may be reduced, exposing platinum to attack by silicon. Therefore, high purity alumina protecting tubes must be used.

- Tungsten-Rhenium

These thermocouples are especially suitable for high temperatures, but they may be used only in vacuum or in inert or reducing atmospheres. The material is brittle, so it is much more difficult to work with than other thermocouple types.

4-2. Thermocouple and Extension Wire Insulation

Positive and negative wires must be insulated from each other so that no current flows between them except at the measuring instrument. Even ceramics, which are considered to be non-conducting at room temperature, can become somewhat conductive at high temperatures, so insulation for use at high temperatures must be selected with care.

The insulation must not interact chemically with the wire. For base metal thermocouples, almost any ceramic may be used. For the platinum alloys, Types R, S and B, silicon must be avoided. For very high temperatures, alumina and beryllia are used with tungsten-rhenium alloy thermocouples.

Extension wire insulation normally must maintain its properties only to 400°F (205°C). The types of insulation that may be used are, therefore, more numerous than those for thermocouple wire. However, in practice the insulation systems offered for extension wire are very similar to those offered for thermocouples, except that the very high temperature fiber insulations such as Refrasil™ are not needed for extension wire.

Table 4-2 shows the temperature limits for use, and the electrical, mechanical, and other pertinent characteristics of common insulation systems. Table 4-2 is not encyclopedic. Many insulation systems combine materials, e.g., fiberglass with silicone over asbestos for better abrasion resistance. The judgments of environmental resistance and the temperature range of use may differ from vendor to vendor.

For temperatures above approximately 1290°F (700°C), most insulation that can be applied directly to a wire is not suitable. Thermocouple wires must be isolated from each other by ceramic insulators. These may be of single- or double-hole design, strung like beads on the wire as shown in Fig. 4-1. Four-hole designs are for duplex thermocouples, two thermocouples in a single tube. Often, one is used for control or indication and the other is used for high limit protection. Figure 4-1 also illustrates that heavy thermocouples may be constructed without insulation if the wires are used at a temperature at which there is no danger of deformation.

Refrasil is a registered trademark of H.I. Thompson Co.

Insulation	Temperature Limits	Electrical Resistance	Moisture Resistance	Abrasion Resistance	Aging Results	Remarks
Nylon	-40 to 160°C -50 to 320°F	Good	Good	Excellent	Excellent	Indoor, light duty use only, not for use in conduits.
Polyvinyl Chloride	-40 to 105°C -40 to 220°F	Excellent	Excellent	Excellent	Excellent	Indoor, outdoor, underground use. Good chemical resistance.
Enamel	to 107°C to 225°F	Good	Fair	Poor	Excellent	For laboratory use only.
Cotton over enamel	to 107°C to 225°F	Good	Fair	Fair	Fair	Not suitable for moist locations
Teflon and Fiberglas	-120 to 250°C -190 to 482°F	Excellent	Excellent	Poor	Excellent	
Fiberglas with Silicone over Asbestos	78 to 482°C -605 to 900°F	Good	Good to 260°C, 500°F	Fair to 260°C, 500°F	Excellent	
Asbestos	-78 to 650°C -105 to 1200°F	Good	Very Poor	Poor	Excellent	Dry atmospheres only.
Tempered Fiberglas	to 650°C to 1200°F	Good	Poor	Fair	Excellent	Dry atmospheres only.
Refrasil	to 1083°C to 2000°F	Good	Very Poor	Very Poor	Excellent	Dry atmospheres only. Stainless Steel braid improves abrasion resistance.
Silicone rubber over fiberglas	-40 to 232°C -40 to 450°F	Good	Good	Fair	Excellent	Food and chemical applications, indoors.

TABLE 4-2. Properties of Thermocouple Insulation Systems

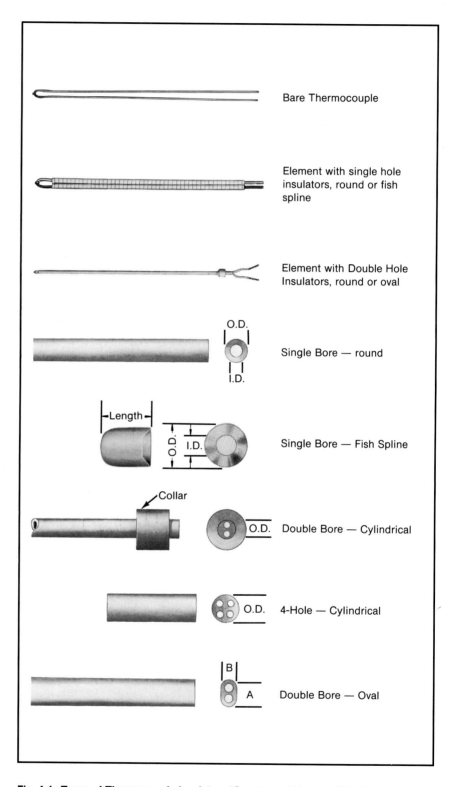

Bare Thermocouple

Element with single hole insulators, round or fish spline

Element with Double Hole Insulators, round or oval

Single Bore — round

Single Bore — Fish Spline

Double Bore — Cylindrical

4-Hole — Cylindrical

Double Bore — Oval

Fig. 4-1. Types of Thermocouple Insulators (Courtesy of Honeywell Inc.)

Table 4-3 shows the standard color codes for the insulation of standardized thermocouples.

Type	Positive	Negative	Overall T/C	Overall Extension Wire
E	Purple	Red	Brown	Purple
J	White	Red	Brown	Black
K	Yellow	Red	Brown	Yellow
T	Blue	Red	Brown	Blue
R	Black	Red	—	Green
S	Black	Red	—	Green
B	Gray	Red	—	Gray

TABLE 4-3. Standard Color Coding of Thermocouples

4-3. Sheathed Thermocouples

Sheathed thermocouples are similar to electric heating elements in construction. Ceramic insulators are strung on wires, the sheath is pulled over the insulated wires, and then the assembly is compacted by swaging, rolling, or drawing.

In swaging, there is relatively little reduction in diameter, so the entire length of the thermocouple must be strung. It is, therefore, relatively costly. This method of manufacture is used where calibration must be closely controlled and material properties must be maintained. Swaging works the material less, so both thermoelectric and material properties are less affected by the manufacturing process.

In rolling or drawing, short lengths of large diameter wire are strung with insulators and placed in a large diameter sheath. The assembly is compacted and reduced in diameter by many passes through drawing or rolling dies, with stress-relieving annealing between passes. Most sheathed T/C material is rolled or drawn because it is cheaper to manufacture. However, the material is strained more than swaged material, so control of thermoelectric properties is more difficult.

Sheathed material can give the thermocouple excellent protection from its mechanical and chemical environment. It can be bent in complex forms at installation and can be welded to the structure or the surface whose temperature is to be measured.

Sheathed thermocouples are essentially monolithic, so they are more difficult to strip, form a junction, and terminate than other constructions.

Thermocouples made from sheathed materials are offered in three hot junction configurations (see Fig. 4-2). In Fig. 4-2 (c), The junction is deliberately insulated with ceramic before the plug is welded to the sheath. The integral junction shown in (a) contains plug and thermocouple wires that may be welded in a single operation so the junction is integral with the sheath. The exposed junction in (b) has the highest speed of response. The sheath is sealed with cement to prevent entry of moisture or gas.

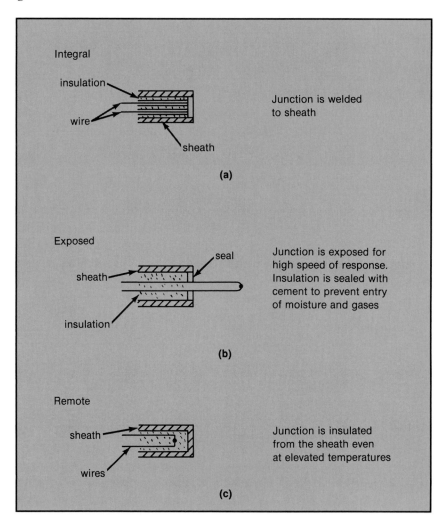

Fig. 4-2. Hot Junction Configurations of Sheathed Thermocouples

Exercises

4-1. *Why should Type J thermocouples not be used in moist atmospheres or at sub-freezing temperatures?*

4-2. *For moist atmospheres what might be a better choice of thermocouple than Type J?*

4-3. *Why are Type K thermocouples normally used only in oxidizing atmospheres?*

4-4. *What is the danger of using silica insulators or protecting tubes with platinum alloy thermocouples?*

4-5. *What is the field of application of tungsten-rhenium types of thermocouples?*

4-6. *Why are most sheathed thermocouples rolled or drawn?*

4-7. *What are the three common hot junction constructions of sheathed thermocouples?*

Unit 5:
Thermocouples in Use

UNIT 5

Thermocouples in Use

This unit presents an overview of the practical aspects of using thermocouples for temperature measurement. First, the elements of typical thermocouple assemblies are reviewed, with comments on the selection of elements such as wells and protecting tubes to meet the requirements of the intended application. A brief discussion of ways to form the measuring junction follows. Specialized constructions for surface temperature measurement and molten metal bath temperatures are discussed.

After this overview, this unit considers the factors involved in selecting a suitable thermocouple design for an application and discusses how the thermocouple installation affects the validity of the temperature reading.

Learning Objectives — **When you have completed this unit you should:**

A. **Have a working knowledge of typical thermocouple assemblies.**

B. **Know how to form a measuring junction.**

C. **Know how to select a suitable thermocouple design for an application.**

5-1. The Thermocouple Assembly

Figure 5-1 shows the elements of a typical thermocouple assembly. Not all elements are necessary for all thermocouple applications. Each element depicted in Fig. 5-1 is representative only. Each illustrates only one of a wide variety of options available. Each application demands decisions on which elements are needed and which kind of element should be selected.

Insulated Thermocouple

A thermocouple may be insulated with ceramic insulation as shown for high temperatures or with insulation applied to the

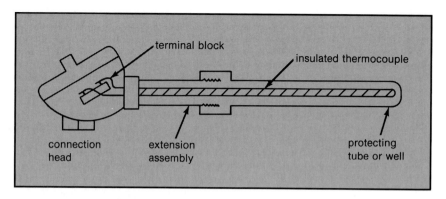

Fig. 5-1. A Typical Thermocouple Assembly

wire itself for moderate or low temperatures. Where large diameter wires are used, there may be no insulation, electrical isolation of the wires being provided by spacing alone.

Protecting Tube or Well

A protecting tube or well protects the thermocouple from contact with the process fluid being measured. It may be a part of the pressure vessel. A protecting tube may be ceramic or metallic. Wells are usually metallic, of robust mechanical construction. Details are shown in Fig. 5-2. A review of protecting tube materials is given in Tables 5-1 and 5-2.

Extension Assembly

An extension assembly is a length of pipe provided between a well or protecting tube and the terminal head to extend the thermocouple assembly to the desired point of measurement, for example, to carry the assembly through a furnace wall or to pierce insulation on a pipe or vessel. An extension assembly is most often used with drilled wells because of the difficulty of drilling long small diameter holes in bar stock.

Terminal Block

A terminal block provides a point for connection of extension wire to the thermocouple assembly. For benign applications it may be a simple block with open terminals, or it may be a quick-disconnect connector.

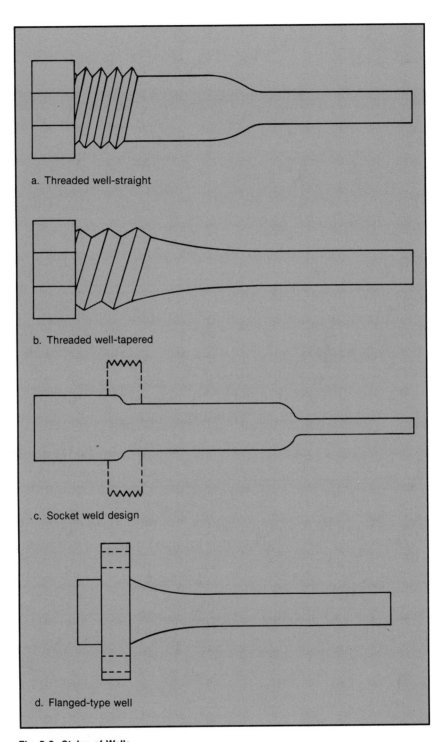

a. Threaded well-straight

b. Threaded well-tapered

.c. Socket weld design

d. Flanged-type well

Fig. 5-2. **Styles of Wells**

Material Name	Composition	Maximum Temperature	Remarks
Quartz	Fused Silica	2300°F/1260°C	Excellent resistance to thermal shock.
Sillramic	Silica-Alumina	3000°F/1650°C	Excellent resistance to thermal shock. Fair mechanical strength. Most frequently used in ovens, furnaces, incinerators, etc. Porous.
Durax	Silicon Carbide	3000°F/1650°C	Gas tight. Often used as inner protection tube inside a Sillramic tube which provides protection against flame impingement.
Mullite	Silica Alumina	3000°F/1650°C	Gas tight. Often used as inner protection tube inside a Sillramic tube which provides protection against flame impingment.
Refrax	Silicon Nitride Bonded Silicon Carbide	3150°F/1735°C	Continuous use in molten aluminum, brass or bronze. Excellent resistance to thermal shock.
High Purity Alumina	99+% alumina	3400°F/1870°C	For use with noble metal thermocouple at above 1200°C. Excellent thermal shock resistance. Gas tight.
LT-1	Chromium-alumina	2600°F/1430°C	Good resistance to mechanical and thermal shock. Excellent resistance to sulphides.

TABLE 5-1. Ceramic Protecting Tube Materials

Connection or Terminal Head

A connection head and cover protect the terminal block from dirt, mechanical damage, and moisture. In benign applications a connection head may not be used. In locations where flammable vapors or gases are present, the head may be of explosion-proof construction. For moderate temperatures the connection head may be plastic.

5-2. Protecting Tubes

For high temperature applications, protecting tubes are usually ceramic, commonly mullite or high purity alumina, although other oxides may be used for special applications. Above 2190°F (1200°C), alumina should be used with noble metal thermocouples because the silicon in mullite will poison the thermocouple, changing its calibration.

For low temperatures, the protecting tube may be a thin metallic tube with the end spun closed and welded or plug welded. For corrosion resistance it may be coated with plastic, lead, or glass.

| Material | Maximum Temperature | | Remarks |
	Reducing Atmosphere	Oxidizing Atmosphere	
Carbon Steel	1000° F/540° C	1000° F/540° C	Noncorrosive liquids and gases only.
Cast Iron	1600° F/820° C	1300° F/700° C	For molten aluminum.
304 Stainless 18% Chrome 8% Nickel	1800° F/980° C	1800° F/980° C	Good resistance to corrosion and oxidation.
316 Stainless 18% Chrome 8% Nickel 2% Molybdenum	1800° F/980° C	1800° F/980° C	Better resistance to corrosion than 304 stainless.
446 Stainless 26% Chrome	1800° F/980° C	1800° F/980° C	For use in sulfurous atmospheres, molten metal and salt baths. Excellent resistance to corrosion and oxidation at high temperatures. Do not use in carburizing atmospheres.
Nickel	1800° F/980° C	1800° F/980° C	Good high temperature corrosion resistance. Excellent high temperature oxidation resistance. Do not use in presence of sulphur above 1000° F (540° C).
Cast-T 35% Nickel 15% Chrome 50% Iron	2000° F/1090° C	2000° F/1090° C	For sulfurous or reducing temperatures.
Inconel 601	2200° F/1205° C	2200° F/1205° C	Good resistance to corrosion and excellent resistance to oxidation at high temperature. Do not use in presence of sulphur above 1000° F (540° C).
Kanthal 22% Chromium 5% Aluminum Balance Iron	2300° F/1260° C	2300° F/1260° C	Good resistance to sulphides in oxidizing atmospheres.

TABLE 5-2. Characteristics of Metal Protecting Tubes

The metal chosen must be compatible with the process fluid into which the thermocouple assembly is immersed. Tables 5-1 and 5-2 list the characteristics of representative protecting tubes.

5-3. Wells

Wells are metallic protecting tubes that are screwed or welded into a pipe or vessel so that they are a part of the process fluid containment system. They may have to withstand high pressure, high temperature, or force from the impact of flowing fluid. They must be compatible with the material of the vessel or pipe for welding and for corrosion resistance. Figure 5-2 shows typical metal wells, illustrating flanged, screwed, and welded constructions.

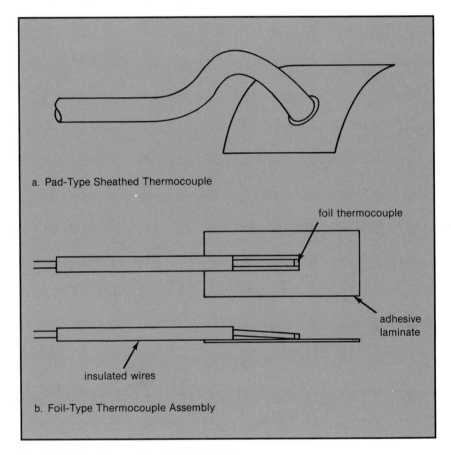

a. Pad-Type Sheathed Thermocouple

foil thermocouple

adhesive laminate

insulated wires

b. Foil-Type Thermocouple Assembly

Fig. 5-3. Surface Temperature Measurement

Wells may be either machined from solid stock for the most demanding service or fabricated by machining and welding for less severe service conditions.

To obtain high speed of response, the thermocouple may be spring loaded so that the hot junction bears firmly against the tip of the well.

5-4. Surface Temperature Measurement

Even though a measurement of the temperature of the gas or liquid in a pipe is desired, it may not be possible to insert a thermocouple assembly into the pipe for fear of eventual pipe blockage or because the pipe is too small. One must then infer the process fluid temperature from the temperature of the pipe surface.

Other applications require the measurement of the temperature of a thin strip of metal or the surface temperature of a piece of equipment.

For the type of application represented by pipe surface temperature measurement, a variety of solutions has been used. In one approach, where the thermocouple may be exposed to high temperatures, a sheathed thermocouple is welded to a metal saddle so that when the saddle is welded or clamped to the pipe the thermocouple junction is at the pipe surface. If the pipe is in a benign atmosphere, the junction of a thermocouple with flexible insulation is peened into the surface of the pipe. The wire is wrapped around the pipe several times to eliminate strain on the joint and to minimize thermal conduction error.

A number of available thermocouple assemblies can be cemented on the surface to be measured. Typically, these consist of thin foil or small diameter thermocouples sandwiched between plastic sheets. They are useful to about 752°F (400°C), depending on the kind of laminating material selected. They may be self-adhering or designed to be affixed with ceramic or organic cements. In some designs, the foil can be removed at installation. Figure 5-3 illustrates some surface temperature measurement assemblies.

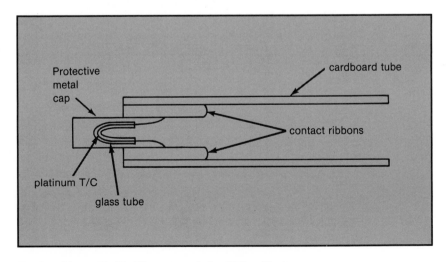

Fig. 5-4. Disposable Tip Thermocouple for Molten Steel

5-5. Disposable Tip Thermocouples

One of the most challenging temperature measurement applications has been the measurement of the temperature of molten steel. It has been possible to continuously measure the temperature of metals with lower melting points, such as brass and aluminum, but no protecting tube material will withstand for long the combination of temperature, mechanical erosion, and chemistry of a molten steel bath. Figure 5-4 shows a solution to this temperature measurement problem. The disposable tip thermocouple assembly is plugged into the end of a lance, which the operator thrusts through the slag layer to the underlying molten metal. The thermocouple junction in a glass tube is protected by a thin metal cap and a cardboard tube. The metal cap protects the thermocouple as the probe passes through the slag, then melts. The glass-enclosed thermocouple reaches metal temperature before it, in turn, melts. The cardboard tube chars but does not disappear before the thermocouple reaches metal temperature.

The measuring system may use a peak picking instrument that remembers the maximum temperature of the thermocouple before it is destroyed, or the history of the temperature during the dip may be recorded.

5-6. Fabricating Thermocouple Hot Junctions

For low temperature work in the laboratory, a hot junction may be formed by simply twisting the wires together, or, for temperatures of a few hundred degrees Fahrenheit, a soft solder may be used. But for continuous measurement in an industrial process it is essential that the junction be formed in a way that it will remain electrically and mechanically intact.

Table 5-3 lists the common methods of soldering or welding hot junctions, and Fig. 5-5 shows four methods of forming junctions in preparation for welding or soldering.

Before gas or arc welding of Types E, J, K, or T thermocouples, the oxide must be removed using abrasive paper. Larger wires should be twisted together for extra strength. Smaller wires may be formed into a "V". Noble metal thermocouples need not be sanded to remove oxides before welding but they must be free of dirt and oil.

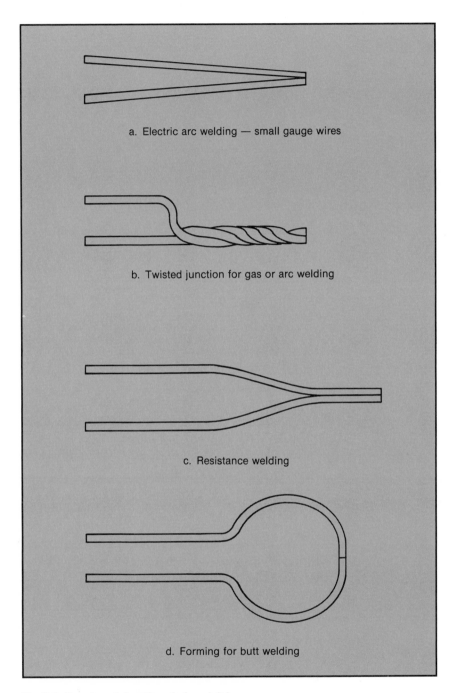

a. Electric arc welding — small gauge wires

b. Twisted junction for gas or arc welding

c. Resistance welding

d. Forming for butt welding

Fig. 5-5. Forming of Junctions before Joining

For gas welding, adjust to a neutral flame. An oxidizing flame may damage the junction. Position the wires in a "V" or twist the wires together after cleaning oxides from the wires (sanding is not necessary for Types B, R, S).

Thermocouple Type	Soft Solder	Silver Solder	Oxygen Acetylene Gas Weld	Electric Arc Weld	Electric Resistance Weld
B, R, S	No	No	No flux	No flux	Yes
K	No	Borax/flux	Borax or fluorite flux	No flux	No flux
J	Resin flux	Borax/flux	Borax or fluorite flux	No flux	No flux
T	Resin flux	Borax/flux	No	No flux	No
E	No	Borax/flux	Borax or fluorite flux	No flux	No

TABLE 5-3. Methods of Joining Wires to Form Hot Junctions

Resistance welding is often used for larger gauge wires. Oxides must be removed before welding. For larger wire sizes, one negative carbon electrode may be used with the wires serving as the positive electrode if dc is used. For smaller sizes, use two carbon electrodes. Form a small solid bead between a "V" junction, or twist the wires. Use a short cycle to avoid burning.

Electric resistance welding is suitable only for larger wire sizes of Type J, K, and B, R, and S. The pressure-time cycle must be established by trial and error.

Other methods include total inert gas and plasma arc methods. These are now often used because no flux is needed. Inert gas shields the junction. This method is especially desirable for tungsten alloy thermocouples.

5-7. Select a Thermocouple for Industrial Temperature Measurement

The following factors must be considered in the process of selecting a thermocouple for a particular industrial temperature measurement application:

- Accuracy required
- Stability
- Reliability and life
- Initial cost
- Replacement cost

When judging all factors except accuracy, the following must be considered:

- Temperature range
- Chemical environment
- Shock and vibration

Table 4-1 shows that accuracy alone is not a significant factor in selecting a thermocouple type.

The first major screen for selection is the temperature to be measured. Between 32°F (0°C) and 662°F (350°C) any thermocouple type may be used. Referring to Table 4-1 one can see that, as the temperature to be measured increases, the range of choice decreases. Above 1832°F (1000°C) only Type K, noble metal, or tungsten alloy T/Cs may be used. Above 3092°F (1700°C) only tungsten alloys may be used.

After the temperature range to be measured has narrowed down the number of types that may be used, the nature of the environment may determine which type must be selected.

Type J should only be used in dry atmospheres. Types B, R, S, E, and K should not be used in reducing atmospheres (locations where oxygen is deficient, as in a sealed thermocouple well). Type T can be used in moist atmospheres, so it can be selected for low or sub-zero temperature measurement.

The tungsten thermocouples are readily oxidized, so they must always be used in an inert atmosphere or in a highly reducing atmosphere. The tungsten alloys tend to be brittle, so they are not suited for locations where the thermocouple may be subject to high vibration or shock.

5-8. Installation Affects Accuracy

When measuring temperature with a thermocouple, we assume that the thermocouple hot junction is at the same temperature as the gas, liquid, or solid whose temperature we want to know. In most situations we don't think much about whether that assumption is true. It is true enough for our purposes. But in some applications it is necessary to be very suspicious and thoughtful about whether the thermocouple temperature is really the product temperature. Heat flows from a hot region to a cooler one by conduction, convection, and radiation. The objective of an accurate thermocouple installation is to ensure that the amount of heat flowing between the point whose temperature is being measured and the thermocouple junction is not sufficient to change the temperature of the measured point or to cause a significant temperature difference between the hot junction and the measured point.

Consider the measurement of the temperature of a stirred liquid in a vat. For practical purposes we can consider the entire volume of liquid to be at the same temperature. If we insert a thermocouple assembly with a half-inch diameter stainless steel protecting tube into the bath, heat flows along the protecting tube towards the colder thermocouple head. If the tube is immersed only one-half inch, we can sense that the thermocouple junction is probably colder than the liquid because of the temperature gradient along or across the protecting tube. As the depth of immersion is increased, the hot junction temperature more nearly equals the liquid temperature, because more of the protecting tube is at the same temperature as the liquid and there is little or no heat flowing in the region of the hot junction. If no heat flows, there is no temperature difference. For this reason it is generally considered that the depth of immersion of a well or protecting tube should be at least 10 times its diameter. And, of course, the mass of the thermocouple assembly should be small compared to the mass of the liquid. One wouldn't think of trying to measure the temperature of coffee in a cup with a half inch diameter tube! Insertion would obviously change the temperature of the coffee.

Surface temperature measurements are also subject to conduction errors. If a hot junction is peened or welded to the surface of a pipe, heat will flow along the thermocouple wire so that there will be a local temperature drop or rise at the point of attachment, depending on whether the thermocouple wire is hotter or cooler than the pipe. The corrective action is to try to arrange the thermocouple so that a significant length is at the same temperature as the hot junction so heat does not flow to or from the hot junction. This is done by wrapping the wire around the pipe or by welding a section of a sheathed assembly near the hot junction to the pipe.

Suppose the temperature to be measured is the temperature of a gas at 2000°F (1093°C) flowing in a duct. The problem of conduction along the thermocouple assembly is similar to that of the liquid measurement application and can be dealt with by deep immersion. However, if the duct walls are cooler than the gas, the thermocouple will read lower than the gas temperature. Why? Because of radiation errors. The thermocouple tries to be at the same temperature as the gas, but it radiates energy to the cool walls and, therefore, drops in temperature. The solution is to provide a shield around the thermocouple that is also exposed to the flowing hot gas. Now the hot junction radiates to the shield. Because they are at nearly the same temperature, the amount of heat radiated is much lower. For some gas measurements, thermocouple assemblies with two or even three concentric shields have been used to minimize radiation errors.

To generalize, then, the thermocouple installation should always be arranged to minimize the flow of heat in the region of the hot junction due to conduction or radiation. If the flow of heat is very small in the region of the hot junction, the temperature gradient will be small and the error will also be small.

Exercises

5-1. *For outdoor installation in a well, would a Type K thermocouple or a Type J thermocouple be a better choice for measuring a temperature of 1110°F (600°C)?*

5-2. *How can spring loading affect the accuracy of a thermocouple?*

5-3. *Why are shielded thermocouples used?*

5-4. *Can a thermocouple be used to measure the temperature of a sheet of metal 0.020 in. thick?*

Unit 6:
Resistance Thermometers

UNIT 6

Resistance Thermometers

This unit discusses the necessary characteristics of any resistance temperature detector and the construction of the most common detectors. It also discusses the operating characteristics of typical platinum resistance temperature detectors (RTD) and how the resistance can be measured and interpreted as temperature.

Learning Objective — When you have completed this unit, you should:

A. Have a working knowledge of resistance temperature detectors and their uses in measuring temperature.

6-1. General Requirements for Resistance Thermometry

A resistance thermometer, to be a useful means to measure temperature, must have a predictable and stable relationship between resistance and temperature. In order that detectors be as small as practicable, the specific resistance of the wire or film must be relatively high so that the resistance can be easily measured. The change in resistance with temperature should also be high. The wire must be mechanically strong, and its resistance-temperature characteristic should not be significantly affected by the strains induced by winding the temperature-sensitive element. And, of course, there must be no change of phase or state over the desired range of temperature measurement. Lastly, the material must be commercially available with a consistent temperature-resistance relationship so that detector calibration is uniform from unit to unit.

The most common materials having these necessary properties are copper, nickel, platinum, and 70% nickel, 30% iron.

6-2. Wire-Wound Platinum Elements

In industrial temperature measurement, platinum wire-wound elements with a resistance of 100 ohms at 32°F (0°C) are now the most common commercially available units.

Only in recent years has there developed an international standard calibration curve. This standard curve can be expressed as:

For the range $-330°F$ to $32°F$ ($-200°C$ to $0°C$),

$$R_t = R_0 (1 + K_1 t + K_2 t^2 + K_3 (t-100) t^3$$

For the range $32°F$ to $480°F$ ($0°C$ to $250°C$),

$$R_t = R_0 (1 + K_1 t + K_2 t^2)$$

where:

R_t = resistance at temperature t
R_0 = resistance at $32°F$ ($0°C$)
t = temperature, $°C$

The constants are:

$$K_1 = 3.90802 \times 10^{-3} \ °C^{-1}$$
$$K_2 = -5.802 \times 10^{-7} \ °C^{-2}$$
$$K_3 = -4.27350 \times 10^{-12} \ °C^{-4}$$

The quotient $\dfrac{R_{100} - R_0}{100 \ R_0}$ is usually designated as α.

For the standard IEC calibration curve, $\alpha = 0.00385$/ohms/ohm/$°C$.

An abbreviated table of resistance vs. temperature is given in Table 6-1.

°C	0	10	20	30	40	50	60	70	80	90	100
-200	18.49										
-100	60.25	56.19	52.11	48.00	43.87	39.71	35.53	31.32	27.08	22.80	18.49
0	100.00	96.09	92.16	88.22	84.27	80.31	76.33	72.33	68.33	64.30	60.25
0	100.00	103.90	107.79	111.67	115.54	119.40	123.24	127.07	130.89	134.70	138.50
100	138.50	142.29	146.06	149.82	153.58	157.31	161.04	164.76	168.86	172.16	175.84
200	175.84	179.51	183.17	186.32	190.45	194.07	197.69	201.29	204.88	208.45	212.02
300	212.02	215.57	219.12	222.65	226.17	229.67	233.17	236.65	240.13	243.59	247.04
400	247.04	250.88	253.90	257.32	260.72	264.11	267.49	270.86	274.22	277.56	280.80
500	280.90	284.22	287.53	290.83	294.11	297.39	300.65	303.91	307.15	310.38	313.59
600	313.59	316.80	319.99	323.18	326.35	329.51	332.66	335.79	338.92	342.03	345.13
700	345.13	348.22	351.30	354.37	357.42	360.47	363.50	366.52	369.53	372.52	375.51
800	375.51	378.48	381.45	384.40	387.34	390.26					

TABLE 6-1. Resistance-Temperature Relationship of 100-ohm Platinum Detector per IEC Pub. 751

Platinum bulbs with ice point resistance of 100 ohms and α = 0.00385 have been common in Europe since before World War II. Most conformed to the German standard DIN 43760. The DIN 43760 curve now is the same as IEC Pub 751, but earlier curves deviated somewhat from the now IEC standard values, especially at subzero temperatures.

Until the availability of the international standard curve, there was no national standardization of platinum detectors in the United States. Laboratory standard elements were wound using a pure platinum with α = 3.92 × 10^{-2}. Some manufacturers of detectors for industrial use maintained calibration curves based on α = 0.00392. Others used 0.00385, and still others used an intermediate value.

For this reason it is essential that, if a bulb is to be replaced or if the instrument measuring or controlling temperatures from the element is to be replaced, the replacement be calibrated to the same curve as the original.

For example, a 100-ohm bulb with α = 0.00392 will have resistance values as follows:

0°C	100.00
100	139.16
200	177.13
300	213.93
400	249.56
500	284.02
600	317.28

If this bulb were replaced by a bulb with α = 0.00385, the instrument reading resistance will indicate a temperature lower than actual.

At 500°C the IEC bulb will have a resistance of 280.90 ohms. The other bulb has a resistance of 280.90 ohms at approximately 490°C. This will be the indicated temperature of the instrument that has been calibrated to the α = 0.00392 curve.

Most manufacturers will sell bulbs to the IEC curve as well as to older curves, and instrument manufacturers are tending to base temperature calibration on the IEC Standard curve.

Even though a bulb calibrated to the IEC Standard curve has α = 0.00385 and bulbs according to DIN 43760, before it was changed to agree with the IEC curve, also had α = 0.00385, the curves are not identical. At $-100°C$ the DIN curve called for R = 60.23 ohms compared to 60.25 for the IEC curve. This is a difference of approximately $0.05°C$. At $800°C$ the difference between the curves is about one-third degree. These differences are not likely to be significant in practice.

Resistance elements for industrial use are manufactured of small diameter wire, 0.001 in. diameter being typical. The wire may be wound in a coil and inserted in a hole in a ceramic insulator, or the coil may be wound on a ceramic or glass bobbin. Usually the element is made monolithic by enclosing it in a fired ceramic frit or by enclosing it in a glass or ceramic capsule. The specific construction depends on the desired usable temperature range. Glass constructions are usable to $300°$ to $500°C$. For higher temperatures ceramic, usually alumina (aluminum oxide), is used for core and insulation.

Elements are available in diameters of 0.06 to 0.18 in. (1.5 to 4.5 mm) and lengths of 1 to 2 in. (25 to 50 mm).

Though elements of 100-ohms resistance at the ice point are most common, elements with $0°C$ resistance of 50 ohms or multiples of 100 ohms, such as 500 and 1000 ohms, are available. Elements are also available with two and three separate windings in a single-bulb assembly. A two or three winding element might be used where it is desirable to have redundant measurements for safety. Or, one winding might be used as the temperature sensor for control, and the second would provide input to a high or low temperature limit controller.

The resistance element is always protected from moisture and the environment by a sheath, usually stainless steel, Inconel™, or some other temperature and corrosion resistant material. A representative assembly is shown in Fig. 6-1. Many variations are possible, depending on the speed of response and maximum temperature desired, but the functions of the parts of the assembly are the same regardless of the specific design.

The element must fit tightly inside the sheath to obtain high rate of heat transfer for fast speed of response. To eliminate air

---

Inconel is a registered trademark of Huntington Alloys, Inc.

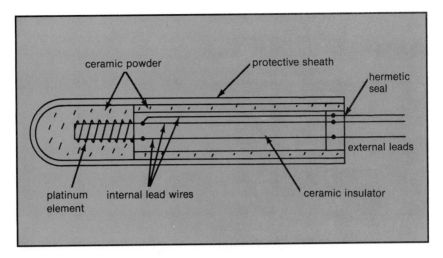

Fig. 6-1. A Representative Resistance Thermometer Probe

pockets, which have low thermal conductance, a fine powder
filling may be used.

The internal lead wires must be suitable for the temperature
range of use, so they are usually nickel or a nickel alloy. They
must be well insulated from each other and from the sheath so
that leakage resistance does not degrade the inherent accuracy
of the element. Ceramic insulators are commonly used.

At the end of the tube a hermetic seal suitable for the usable
temperature range of the probe protects the element, the internal lead wires, and the junction between the internal and
external lead wires.

The probe assembly may be terminated in the external lead
wires, or it may be supplied with a ceramic terminal block,
cast iron, or aluminum connection head, or any of the varieties
of connection facilities that might be supplied in a thermocouple assembly.

Resistance thermometer probes are available with 2, 3, or 4
internal and external leads depending on the measurement system being used. The various ways to measure the resistance of
an RTD are discussed in Section 6-7.

If a temperature measurement is to be made in a vessel with
high internal pressure, a pipe with high fluid velocity, or in the
presence of corrosive liquids, the temperature probe may be
further protected by insertion in a well or protecting tube of a
design similar to those used for thermocouples.

To increase speed of response of a probe inserted in a well, spring-loaded assemblies are available in which the end of the probe is held against the bottom of the well to ensure good thermal contact. In some cases, the probe may be provided with a highly conductive tip of silver to further increase thermal response.

6-3. Thin-Film Platinum Elements

Thin-film platinum elements now compete in many ways with wire-wound elements. Prices are competitive with wire-wound elements and are, in many cases, lower because the deposition of the film and trimming to the desired calibration curve can be automated, eliminating the relatively high labor content of wound bulbs.

Thin-film techniques are especially effective for high resistance bulbs, 1000 ohms being a common value to eliminate lead resistance errors. The platinum content of a thin-film element is much lower than in a wire-wound element of the same resistance.

Thin-film bulbs are available in cylindrical constructions so that probe construction is similar to that already described. Cylindrical elements can be fabricated in resistances up to 1000 ohms in sizes smaller than wound bulbs, e.g., 0.08 to 0.12 in. (2 to 3 mm) diameter by 0.2 to 0.5 in. (5 to 12 mm) long.

A significant advantage of thin-film elements is that they can be made in much smaller sizes than wound elements and in flat packages suitable for surface temperature and gas temperature measurements.

Thin-film sensors are now available to match or nearly match the Standard IEC calibration curve. Some may meet the earlier DIN 43760 curve, but in practice the differences between the two curves are unimportant in industrial temperature measurement.

6-4. Nickel Resistance Thermometers

Although nickel resistance thermometers were the thermometers of choice for several decades, they are now of importance primarily for replacement of bulbs in existing measurement and control systems.

Nickel was an attractive metal for resistance thermometer elements because its temperature coefficient of resistance in the pure state is 0.0066 ohms/ohm/°C compared to 0.00393 ohms/ohm/°C for platinum. The price of nickel compared to platinum was also attractive.

However, unlike platinum, which is commercially available with a predictable and reproducible temperature coefficient of resistance, nickel in commercial lots varies considerably in temperature coefficient. It was therefore necessary to provide a series padding resistance of material with a low temperature coefficient of resistivity. By adjusting the resistance of both the nickel element and the series padding resistor, a standard calibration curve could be matched. The addition of a series connected padding resistor, which is not temperature sensitive, reduces the overall temperature coefficient of resistance so that the effective coefficient is not significantly higher than that of a platinum element.

No true standard for nickel elements was ever developed. Each manufacturer had a proprietary curve. Nickel elements are more nonlinear than platinum elements, so some manufacturers devised ways of padding elements to make them more linear.

Nickel element calibration was, in general, based on an ice point value of resistance of several hundred ohms.

Nickel elements were devised at a time when the most common measuring instrument was a self-balancing Wheatstone bridge with D'Arsonval galvanometer error detectors and, later, vacuum tube amplifiers. High resistance elements, typically 300-500 ohms at the ice point, were used to minimize lead resistance errors and maximize the resistance change per degree.

Nickel elements were used primarily in the temperature range up to 300°F (150°C).

6-5. Copper Resistance Elements

Electrolytic copper is commercially available with consistently high purity. Its temperature coefficient of resistance is 0.00425 ohms/ohm/°C, slightly higher than platinum. The specific resistance of copper is low, however, so copper elements of convenient size are limited to 10 to 50 ohms at the ice point. They can be used over the range –330°F to 300°F (–200°C to 150°C). From –60°F to 300°F (–50°C to 150°C) they are sufficiently linear that two sensors can be used directly to measure temperature differences. The most common application of copper bulbs is in measuring the stator temperature in rotating machines.

6-6. Characteristics of Resistance Thermometers

Accuracy

IEC Pub. 751 defines two classes of accuracy for platinum resistance thermometer elements. Class A sensors are accurate to within $\pm$ (0.15 + 0.002$|t|$) °C. Class B sensors are accurate to within $\pm$ (0.3 + 0.005$|t|$) °C. $|t|$ is the temperature without regard to sign.

Commercial sensors generally have an accuracy similar to IEC Class B standards. The typical tolerance at the ice point for a 100-ohm bulb is 0.1 ohm.

Thin-film sensors are available with the same accuracy specifications as wire-wound sensors.

Accuracy of a resistance thermometer sensor can be checked either by comparison with a standard reference element with known calibration or by measuring the resistance at fixed points. The first method is most convenient for most industrial purposes, requiring only that both elements be at the same temperature. Resistance shall be measured using a current that will not cause self-heating of more than 1/5 of the accuracy specification.

Stability

Unlike thermocouples, resistance elements are sealed so that they are not exposed to the atmosphere in which temperature is being measured. Drift of the element's calibration is therefore caused either by contaminants sealed into the capsule during manufacture or by strains that develop as the element is temperature cycled because the temperature coefficient of expansion of the wire or film is not identical to that of the mandrel or substrate on which the metal is wound or deposited.

IEC 751 specifies a test for effects of temperature cycling that consists of 10 cycles from room temperature to the highest rated operating temperature, return to room temperature, then to the lowest rated operating temperature and return to room temperature. The 0°C resistance value shall not change more than the equivalent of 0.15°C for Class A sensors and 0.3°C for Class B sensors. IEC 751 also specifies a test for short-term stability. After 250 hours at the rated temperature extremes, the ice point shall not shift more than 0.15°C for Class A or 0.3°C for Class B sensors.

One manufacturer specifies a 0.05-ohm (1.25°C) change in 0°C resistance after 2000 hours at 600°C. This same manufacturer specifies a 0.02-ohm change in R_0 after cycling per the IEC standard.

Speed of Response

Because the resistance element is inherently more massive than a bare thermocouple, response to temperature changes is slower.

A speed of response specification must specify the medium in which speed is being measured and the flow velocity of the medium. Often, the time constant (the time to measure 63% of a step change) is specified. Some manufacturers specify 50% and 90% responses. The 90% response is a little more than two time constants.

Response times of probes in water moving at 3 ft/second are typically several seconds, faster for small diameter probes (less than 1/4 in. diameter), and slower for larger diameter probes. Bare elements, without a sheath, may have time constants of less than 1 second in water. Response times in air are 50-100 times longer than in water.

Self-Heating

The measurement of the resistance of a resistance thermometer detector demands that a current be passed through the resistance element. This current produces heat that raises the temperature of the element and, therefore, its resistance. The self-heating error is the amount of resistance change, converted to °C, and may be stated by the manufacturer in °C/mw, or mw/°C.

The magnitude of the self-heating error depends on the efficiency of heat transfer from the sensing element to the protective sheath and from the sheath to the medium being measured. The self-heating error is, therefore, much larger when the detector is measuring moving air than when it is measuring moving liquid.

The standard method of determining the self-heating error is to immerse the thermometer in a stirred constant temperature bath, usually an ice bath. Measure the resistance of the bulb at two levels of current and calculate the wattage dissipated at each level of current. The self-heating error is then:

$$\text{Error} = \frac{1}{S} \frac{(R_2 - R_1)}{(W_2 - W_1)}$$

where S is average slope of the calibration curve in ohms/°C at the temperature at which the test is carried out; R_1 and R_2 are the resistances; and W_1 and W_2 are the wattages, respectively, at the two levels of current.

Typical values of error for wound detectors are 0.05 to 0.35°C/mw in air with 1 meter/second velocity. Values in still air are 2 to 4 times as large, depending on detector construction.

In still air, thin-film detectors may have self-heating errors of about 0.5°C/mw. In still water, the self-heating error of representative detectors is about 0.01° to 0.02°C/mw.

Vibration Resistance

Modern industrial resistance temperature detectors have good resistance to vibration. There is no standard specification for vibration resistance. A complete specification would state the maximum change in resistance, expressed in °C calibration change, after being subjected to vibrations over a range of frequencies at a specified acceleration for a stated time.

Manufacturers' literature may not specify vibration resistance at all or may state only a range of frequencies at some acceleration.

Most applications do not subject a resistance temperature detector to high levels of vibration. Applications in pipe lines or diesel engine exhaust systems are noted for high levels of vibration. For such applications, special designs are available with higher vibration resistance than is found in standard industrial detectors.

If the detector is used in a well, damage to the detector often occurs because it vibrates inside the well. Spring loading the detector can reduce the relative motion between the detector and well.

Insulation Resistance

Ideally, the resistance between the sensitive resistance element or lead wires and the protective sheath is infinite. Insulation resistance of at least 100 megohms at room temperature is a typical value, decreasing by a factor of 10 or more at 600°C.

Lower insulation resistance is usually the result of moisture that has invaded the assembly through a flaw in the protective sheath or at the lead wire seal.

At 500°C the standard IEC detector has a resistance of 280.90 ohms. A change of 1°C produces a change of 0.33 ohms, or 0.118% change in resistance. A 0.24-megohm leakage resistance shunting the bulb would produce an error of 1°C at 500°C. A 2.4-megohm leakage would produce a 0.1°C error.

IEC Pub. 751 requires minimum insulation resistance between each terminal and the sheath to be 2 megohms at 500°C.

Immersion Error

The resistance temperature detector, like a thermocouple assembly, will conduct heat from the region in which the temperature is to be measured. However, the resistance element is typically much larger than the hot junction of the thermocouple. Therefore, a rule of thumb for immersion of a resistance thermometer detector is to immerse it a distance equal to the length of the sensitive element plus six times the diameter of the sensitive element.

The minimum depth of immersion can be found by immersing a detector in a well-stirred, heated bath and decreasing the depth of immersion until the measured resistance decreases the equivalent of 0.1°C. This immersion is the minimum that should be used.

6-7. Measurement Methods

At 0°C the resistance of most industrial platinum bulbs is 100 ohms. A change of 1°C produces a change of 0.39 ohm. 18-gauge copper wire has a resistance of approximately 0.0064 ohm per foot. If the measuring instrument is 50 feet from the detector there will be 100 feet of lead, or 0.64 ohm, equivalent to an error of 1.64°C.

Most industrial thermometers are provided with three leads. If the instrument for measuring the resistance of the detector is a Wheatstone bridge, the circuit is:

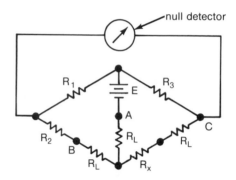

A, B, and C are the connections to the resistance thermometer bulb.

If, as has usually been the case, the value of R_2 is changed to maintain zero voltage across the error detector, the effects of lead resistance are reduced to zero.

At balance:

$$\frac{R_3}{R_1} = \frac{R_x + R_L}{R_2 + R_L}$$

The three lead resistances R_L are assumed to be the same, as in practice they are. One resistance is in series with the bridge supply and has no significant effect on accuracy. The other two lead resistances are in series with R_x, the resistance of the sensitive element, and R_2, the variable rebalancing resistance. But at balance, $R_x = R_2$.

Modern digital electronic measuring devices can now use the three-lead bulb in a different way.

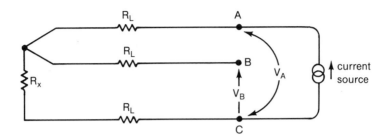

A constant current source drives a current through $R_x + 2R_L$. A voltage detector reads a voltage proportional to $R_x + R_L$ at B and a voltage proportional to $R_x + 2R_L$ at A. $V_A - V_B$ is proportional to R_L. So $V_B - V_A + V_B$ is proportional to R_x alone. The measurements are made sequentially, digitized, and stored until the differences can be computed.

This measurement method assumes, as does the balanced Wheatstone bridge approach, that the lead resistance is the same in all three leads and small relative to the bulb resistance. This is normally so, for practical purposes, unless lead lengths are very long. In such cases, transmitters that convert the resistance measurement to a 4-20 ma signal may be used to allow short leads to be used to connect the resistance temperature detector to the transmitter.

For the highest level of accuracy, four-lead bulbs are used. These are of two types.

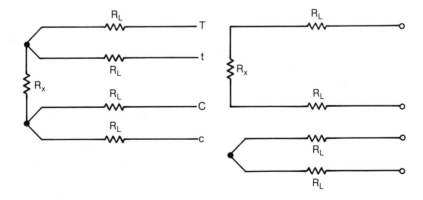

The four-lead bulb shown in (a) has been commonly used in high precision laboratory measurements.

The first measurement is made in a bridge configuration as shown below.

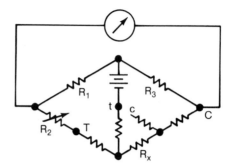

The sensor connections are then changed to:

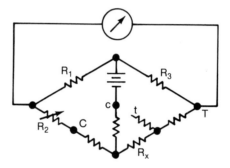

The two measurements of resistance are averaged and are independent of lead resistance.

Today, this four-wire bulb can be used in the circuit below.

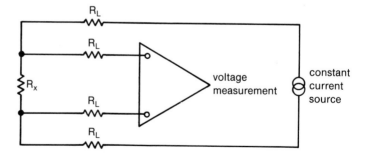

In this use of a four-lead bulb, the constant current source provides a current through R_x, which is essentially independent of the values of R_x and R_L.

The voltage measurement amplifier has high enough impedance so that variations in R_L do not affect the voltage
measurement.

The second four-lead configuration illustrated above (b) is not
commonly used in North America. It is available commercially
and is recognized in IEC Pub. 751. By measuring the resistance
$R_x + 2R_L$ and the resistance $2R_L$, a measurement of R_x can be
computed.

In analog instruments one can visualize a circuit such as shown
below:

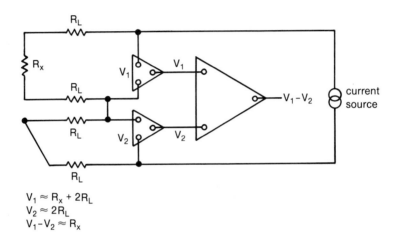

$$V_1 \approx R_x + 2R_L$$
$$V_2 \approx 2R_L$$
$$V_1 - V_2 \approx R_x$$

Exercises

6-1.   *Why are copper resistance bulbs not used in industry as
       commonly as platinum bulbs?*

6-2.   *Why are nickel bulbs no longer popular?*

6-3.   *Although most industrial measurements of resistance
       bulbs use three leads, under what conditions might two
       leads be satisfactory?*

6-4.   *For what reasons are resistance bulbs spring loaded?*

6-5.   *The measuring current supplied to a resistance bulb by
       a controller is 0.5 ma. What error would this cause in a
       bulb if the bulb manufacturer states a self-heating error
       to the condition of use of 0.1°C/mw?*

6-6.   *How would the effects of leakage resistance be apparent in the operation of a bulb?*

6-7.   *How can one determine if a resistance bulb is immersed in a fluid to a sufficient depth to obtain an accurate measurement?*

Unit 7:
Filled Systems

UNIT 7

Filled Systems

This unit discusses the construction and principles of operation of the common types of filled-system thermometers and compares their performance.

Learning Objective — When you have completed this unit you should:

A. Have a working knowledge of the operations and application limits of filled thermometer systems.

7-1. Principle of Operation

The filled-system thermometer consists of four parts as illustrated in Fig. 7-1.

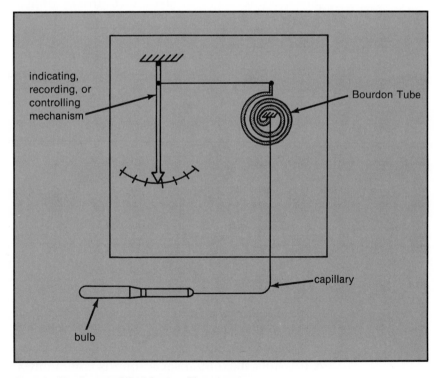

Fig. 7-1. The Generic Filled System Thermometer

A capillary tube connects a bulb containing a fluid that is sensitive to temperature changes to an element that is sensitive to
pressure or volume changes. The pressure-sensitive or volume-
sensitive element may be a Bourdon tube, as illustrated, or a
helix, diaphragm, or bellows. (A helix operates in the same
fashion as a Bourdon, but unlike the Bourdon, whose turns lie
in the same plane, the turns of a helix form a cylindrical
surface.)

The motion of the temperature-sensitive or volume-sensitive
element is coupled mechanically to the indicating, recording, or
controlling device.

The following are advantages of filled system thermometers:

- Simple, rugged construction, requiring little maintenance.

- No auxiliary power is required unless to drive a chart or to
  power a pneumatic or electric control or transmission
  system.

- Filled systems can develop significant power to perform
  recording or control functions or, in some applications, to
  operate valves.

The following are major limitations of filled systems:

- The bulbs may be inconveniently large.

- The characteristics of filled systems vary significantly with
  the type of fill fluid, and a user must be careful not to mis-
  apply a system.

- Failure of any part of the system requires that the entire
  filled system be replaced.

- The permissible separation between the bulb and the Bourdon is limited by the characteristics of the fill fluid and the
  accuracy desired, but this typically does not exceed 200 feet.
  A fifty-foot maximum separation is commonly specified.

Filled systems are being replaced in industrial process control
applications by thermocouples and resistance thermometers.
The low cost of electronic devices to read the output of thermocouples and resistance thermometers and to indicate or con-

trol, together with the ability to locate the sensor remote from the receiver and to replace the sensor independently of the receiving device, has made electronic means more attractive. However, filled systems are still common for temperature measurement and control in appliances and similar applications. Simple temperature controls in ovens, refrigerators, air conditioners, and in some heating systems take advantage of the simplicity, low cost, and reliability of filled systems.

7-2. Classification of Filled Systems

Liquid-filled systems are systems in which the bulb, the capillary, and the Bourdon are completely filled with a liquid. An increase in temperature increases the volume of the fill liquid more than the volume of the metallic parts. The increase in fill fluid volume deflects the Bourdon tube.

Vapor- or gas-filled systems are filled with a gas or are partially filled with volatile liquid. An increase of temperature increases the pressure of the gas or vapor and deflects the Bourdon tube.

The Scientific Apparatus Makers Association established a classification system in their Standard, SAMA: RC6-10-1963.

For liquid-filled systems:

Class I — Liquids other than mercury
Class V — Mercury filled

For gas- and vapor-filled systems:

Class II — Vapor-filled systems
Class III — Gas-filled systems

7-3. Liquid-Filled Systems

An ideal temperature measurement system should respond only to the temperature at the sensing element. In liquid-filled systems, however, because the temperature-sensitive fluid completely fills the bulb, capillary, and Bourdon, a change in the temperature anywhere in the system changes the fill fluid volume. If the bulb temperature is held constant, a change in ambient temperature along the capillary or at the Bourdon tube will cause the indicated temperature to change.

The simplest means to compensate for this error assumes that the capillary and Bourdon tube temperatures change together. A bimetallic element is inserted between the Bourdon tube and the indicating mechanism to compensate for the error. This is called case compensation.

The most accurate compensation, which does not assume that the capillary and case temperatures change together, is called full compensation. In full compensation an auxiliary thermal system, with a capillary and a Bourdon tube identical to that of the primary system but without a bulb, provides a motion dependent on the temperature along the two capillaries, which are run side to side. The motion of the auxiliary thermal system subtracts from the motion of the primary system, so the indication at the instrument is caused, essentially, only by the temperature of the filled system bulb.

In the SAMA classification system, case-compensated liquid-filled systems are designated IB and VB. Fully compensated systems are designated IA and VA. In Type VA systems it is possible to achieve fully compensated performance without providing a separate compensating Bourdon tube and capillary. Nickel steel wire with a low temperature coefficient of expansion is inserted into the capillary. The dimensions are chosen so that temperature changes along the capillary cause the mercury fill and the capillary volumes to increase by the same amount. The Bourdon tube sees no net change in fill volume.

In mercury-filled systems, if the bulb is designed to be operated over a wide temperature range, the errors caused by changes of temperature along the capillary may be acceptably small, because the volume of the mercury in the capillary is relatively small compared to the volume in the bulb. In such designs, no additional compensation is needed.

Range of Application of Filled Systems

A liquid-filled system must operate above the freezing point of the fill fluid and below the point at which the liquid is unstable or begins to expand in an unacceptably nonlinear fashion.

Many manufacturers design all filled systems to use the same size Bourdon tube, i.e., the same change in system volume is required for 100% deflection of the indicator regardless of the indicated temperature range.

For a given fill fluid, narrowing the temperature range of operation of the bulb requires increasing the size of the bulb.

In general, organic liquids used in filled systems have volumetric coefficients of expansion about 8 times that of mercury. Temperature spans of 10–25°C are obtainable with these fluids, limited by the maximum practical bulb sizes.

Organic fluids are limited to maximum temperatures of 300° to 480°F (150° to 250°C). Mercury-filled systems can be used to 1200°F (650°C).

Elevation Error

If the bulb in a liquid-filled system is elevated above the Bourdon tube or mounted below it, there will be a pressure difference between the bulb and the Bourdon tube, which causes an error in temperature indication. The size of the error depends on the difference in elevation and the density of the fill fluid. For elevation differences of up to 30 feet, head effects are not significant in mercury-filled systems. In other liquid-filled systems they must be compensated by design. The manufacturer will take into account the elevation difference stated by the purchaser when designing the system.

Barometric Errors

Liquid-filled systems are unaffected by barometric pressure changes.

Overrange Protection

Liquid-filled systems can generally withstand temperatures of 150–200% of the upper range value of temperature measurement. A bulb designed to operate at 500°F (260°C) may, for example, be operated at 750°F (400°C) without permanent damage to the system. The manufacturer's specified limits should be observed.

7-4. Vapor Pressure Thermometers

Vapor pressure thermometers are partially filled with a volatile liquid. Temperature change at the bulb changes the vapor pressure in the system, which is translated to motion by the Bourdon tube.

SAMA has defined the following four classes of vapor pressure thermometers, depending on the operating temperature of the bulb relative to the temperature of the rest of the thermal system.

- IIA — The measured temperature is above the temperature of the rest of the system.

- IIB — The measured temperature is below the temperature of the rest of the system.

- IIC — The measured temperature may be above or below but not at the temperature of the rest of the system.

- IID — The measured temperature may be above, below, or at the temperature of the rest of the system.

The construction differences among the four classes of vapor-filled systems arise from two facts. The vapor-liquid interface must always be in the measuring bulb, and the fill will always be liquid in the coolest portion of the system.

In simple terms, the construction is as follows:

- Class IIA — The capillary and Bourdon are liquid-filled. The bulb must contain enough volume to accommodate changes in the Bourdon tube volume over its operating range, changes in the capillary volume caused by temperature changes, and changes in the volume of vapor in the bulb over the operating temperature range. For capillary lengths above 50–75 ft, the bulb size may have to be relatively large.

- Class IIB — This system has the smallest bulb because the capillary and Bourdon tube are filled with vapor, and the bulb is only partially filled with liquid.

- Class IIC — In this system, if the bulb temperature is lower than that of the capillary and Bourdon tube, all the liquid is in the bulb. If the bulb temperature is above the temperature of the capillary and Bourdon tube, the liquid is in the capillary and the Bourdon tube, and the bulb is filled with vapor. The bulb, therefore, has to be large enough to hold the fill volume of the capillary and Bourdon tube. A Class IIC system does not indicate accurately when the bulb temperature is the same as the temperature of the capillary and Bourdon tube. As the two temperatures approach each other, liquid either enters or leaves the capillary and Bourdon tube. This error is sometimes called the "cross-ambient" error.

- Class IID — The bulb in this system is able to operate above, below, or at the temperature of the capillary and Bourdon tube. The capillary and Bourdon tube are filled with a nonvolatile liquid. All the temperature-sensitive volatile liquid is contained in the bulb. The bulb in this system tends to be large, because it must contain enough non-volatile fluid to ensure that the volatile actuating liquid never enters the capillary within the design temperature range.

Range of Application

Vapor-filled systems are generally limited to a temperature range of −40 to 600°F (−40°C to 320°C).

Temperature Compensation

The capillary in a Class II system is not sensitive to ambient temperature changes. In a Class IIA system, the liquid-filled capillary must not exceed the bulb temperature. Increasing ambient temperature changes the modules of elasticity of the Bourdon tube, but this effect is usually small enough that no compensation is provided.

Elevation Error

The pressure in the Bourdon tube equals the vapor pressure of the volatile liquid plus or minus the head of the liquid in the capillary. Therefore, if the bulb is to be at an elevation significantly different than that of the Bourdon tube, the manufacturer must take this into account.

Class IIC systems should not be used where the bulb is at an elevation different from the Bourdon tube, because the capillary can either be filled with liquid or with vapor. No means exist to compensate for the elevation error in both situations.

Barometric Error

In order to minimize the effect of changes in barometric pressure, the pressure in a Class II system is typically 100 pounds per square inch (psi) or more throughout the operating range. Typically, the barometric pressure error is approximately 1/2%.

Overrange Protection

The rate of change of vapor pressure with temperature increases as the temperature increases. Therefore, the degree of over-range protection is less than that of other filled systems.

7-5. Gas-Filled Systems

Gas-filled systems operate on the principle of Charles' Law:

$$\frac{PV}{T} = K$$

The pressures in the bulb, capillary, and Bourdon tube are equal. Therefore:

$$P = K \left(\frac{T_b}{V_b} + \frac{T_c}{V_c} + \frac{T_B}{V_B} \right)$$

where T_b, T_c, and T_B are the absolute temperatures of the bulb, capillary and Bourdon tube, respectively; and

V_b, V_c, and V_B are the volumes of the bulb, capillary, and Bourdon tube, respectively.

Range of Application

The range of application of gas-filled systems is the widest of all types of filled systems. The lowest temperature is limited by the temperature at which the fill becomes liquid. The maximum temperature is determined by either the maximum permissible bulb temperature or the maximum permissible Bourdon tube pressure.

Temperature Compensation

Both full compensation, Class IIIA, and case compensation, Class IIIB, are used. The methods of compensation are the same as for Class II and Class V systems.

Elevation Error

The density of gas is so low that gas-filled systems have no appreciable elevation error.

Barometric Error

To minimize barometric errors, gas-filled systems are usually designed to operate at 400 psi or higher.

Overrange Protection

Gas-filled systems are limited by the characteristics of the bulb and the Bourdon tube, so they can provide significantly greater overrange protection than other types of systems.

7-6. Speed of Response of Filled Systems

The speed of response of a filled system depends on so many factors that it is not possible to generalize. The type of fill, capillary length, and bulb size all affect the speed of response. Typical bulb sizes are 4 to 5 inches long by 3/8 to 3/4 inch in diameter.

Without a well in well-stirred water, response time may typically be 2–10 seconds for 63% of the response for bulbs of the sizes noted above. In wells, the response time will be 10 to 15 times greater.

7-7. Application of Filled Systems

Like thermocouples and resistance thermometers, the bulbs of filled systems may be immersed directly in benign fluids. For corrosion protection or if the fluid is in a closed system, thermo-wells are used.

Although filled systems for industrial use once were much less expensive than a thermocouple and its electronic measuring instrument, the electronic state of the art has eliminated this advantage, except for those applications where the filled system has the advantage of requiring no external power source.

Exercises

7-1.   What are the four classes of filled systems?

7-2.   What is the difference between case-compensated systems and fully compensated systems?

7-3.   When must differences between the elevation of the bulb and that of the instrument be a matter of concern?

7-4.   Why have electronic instruments that read thermocouples or resistance bulbs replaced filled systems in most industrial applications?

Unit 8:
Radiation Thermometers — Theory and Construction

UNIT 8

Radiation Thermometers — Theory and Construction

This unit presents the theory underlying the use of radiation thermometers (also commonly referred to as radiation pyrometers) to measure temperature. This unit also reviews the wide variety of pyrometer constructions that are available. Completion of this unit prepares the student for study of the selection and application of radiation thermometers in Unit 9.

Learning Objectives — When you have completed this unit you should:

A. Understand the theories behind and the construction of radiation thermometers (radiation pyrometers).

B. Be prepared for the study of selection and application of these instruments.

8-1. Why Use Radiation Thermometry?

Radiation thermometers measure the energy radiated from the object whose temperature is being measured. The radiation thermometer can measure this radiation from a distance. There need be no contact between the thermometer and the object because, unlike thermocouples, RTD's, and filled system bulbs, the radiation thermometer need not be at the same temperature as the object being measured.

Radiation thermometers are suited especially to the measurement of moving objects or objects inside vacuum or pressure vessels.

The speed of response of slower types of radiation thermometers is only a few seconds. Many respond much faster.

There is no free lunch. The benefits of radiation thermometers are obtained at a price. Even the simplest radiation thermometer is more costly than a thermocouple or RTD assembly. The cost of installing sighting tubes usually exceeds the cost of installing wells or protecting tubes.

Although the simplest radiation thermometers are rugged and may be in service for decades, they require routine maintenance to keep the sighting path clear and to keep optical elements clean. The sophisticated radiation thermometers needed for more difficult problems of temperature measurement have more complicated optics, some with rotating or moving parts, and usually have electronic signal conditioners, which lower reliability.

Unlike thermocouples and resistance thermometers, radiation thermometers have no industry-wide calibration curves.

Lastly, the advantageous results of using radiation thermometers to solve difficult temperature measurements frequently are obtained only after significant engineering investigation of the particular application to select the optimum radiation thermometer and to install it in a way that yields a reliable measurement of product temperature.

8-2. Types of Radiation Thermometers

The following is a convenient classification of radiation thermometers:

- Broadband thermometers
- Band-pass thermometers
- Narrow band thermometers
- Ratio thermometers
- Optical pyrometers
- Fiber optics thermometers

These classifications are not precise. Optical pyrometers can be considered a subset of narrow band thermometers; and fiber optics thermometers can, in principle, be wide band, band-pass, narrow band, or ratio devices. At one time the terms "wide band" and "band-pass" usually implied relatively simple construction. Now, however, even wide band and band-pass thermometers may have sophisticated optics and electronic signal conditioning.

Broadband Thermometers

Broadband thermometers have usually been the simplest radiation thermometers, with response from 0.3-micron wavelength to an upper limit of 2.5 to 20 microns (μm), determined by the lens or window material. They have been termed "total radiation" thermometers because in the temperature ranges of normal use they measure a significant fraction of all the thermal radiation emitted by the object whose temperature is being measured. A typical broadband pyrometer is shown in Fig. 8-1.

Fig. 8-1. Typical Broadband Pyrometer (Courtesy of Honeywell Inc.)

Band-Pass Thermometers

Band-pass thermometers were initially derived from simple broadband thermometers. Lens, window, or filter characteristics were selected to view only a selected portion of the spectrum. The 5–7 μm band-pass was used to measure glass temperature because glass emits strongly in this region but poorly outside this band.

Many of the early band-pass thermometers used simple optics and thermopile detectors. Now that more advanced detectors and signal conditioning electronics are available, the band-pass technique is uncommon, although at least one manufacturer provides a portable radiation thermometer with 8–14 μm response for general-purpose use.

Narrow Band Thermometers

Narrow band thermometers operate over a narrow range of wavelengths. The spectral response of many narrow band thermometers is determined by the detector used. Such narrow band thermometers are used as general-purpose instruments over the temperature range of interest. For example, a thermometer using a silicon cell detector will have a response that peaks at 0.9 μm. The upper limit of usefulness is about 1.1 μm. Such a thermometer is useful at temperatures above 1100°F (600°C).

Other narrow band thermometers utilize filters to restrict response to a selected wavelength to meet the needs of a particular application. Some examples are:

Wavelength - μm	Application
0.65	Accurate measurement of high temperature objects in the open
3.43 ± 0.2	Thin film, polyethylene-type plastics
5.0 ± 0.2	Glass surface temperature

Ratio Thermometers

A ratio thermometer measures radiated energy in two narrow bands and calculates the ratio of the two energies. This ratio is a temperature-dependent function. The temperature measurement is not primarily dependent on the energy in the two bands, only on the ratio of the two energies. Therefore, any influence that affects the amount of energy in each band by the same percentage has no effect on the temperature indication.

Changes in target size have no effect. If the emissivity of the target is the same at both wavelengths, the indicated temperature is not changed by changes in emissivity. Unfortunately, this condition is not fulfilled by most oxidizable materials. For other materials, the ratio technique may reduce or eliminate changes in indicated temperature caused by changes in surface finish. The ratio technique may reduce the effect of energy-absorbing materials such as water vapor or CO_2 between the target and the thermometer, if the percentage of absorption is the same at each wavelength.

Optical Pyrometers

Although optical pyrometers can be considered a subset of narrow band thermometers, they are, in their most popular form, sufficiently unique in design and use to warrant a separate discussion.

An optical pyrometer measures the radiation from the target in a narrow band of visible wavelengths, centered at about 0.65 micron in the red portion of the spectrum. The most commonly used optical pyrometers are manually operated. The operator sights the pyrometer on the target. At the same time he can see in the eyepiece the image of an internal tungsten lamp filament. When the target image and the tungsten filament image are the same color, the target temperature can be inferred from the tungsten filament temperature. When the target and filament colors are the same, the filament image apparently vanishes, so these pyrometers have also been called disappearing filament pyrometers.

In one optical pyrometer, the operator adjusts the current through the tungsten filament to match the color or brightness of the two images. Because the relationship between filament current and filament temperature is known and is stable, when a color match is achieved the amount of filament current can be interpreted as the temperature of the filament, and, therefore, of the target temperature.

Another form of optical pyrometer maintains a constant current through the filament and adjusts the brightness of the target by means of a rotatable energy-absorbing optical wedge. The amount of energy absorbed by the wedge depends on its angular position. When the color or brightness of the target image and the filament image are the same, the target temperature can be read from the calibrated knob that adjusts the optical wedge.

These manually operated brightness pyrometers have been used for decades as the working reference standard for spot checking whether a radiation thermometer is indicating the correct temperature. Because of the narrow band of wavelengths centered in the visible red portion of the spectrum, the optical pyrometer is much less affected by changes in the properties of the target material and surface condition, and by CO_2 and water vapor in the sighting path, than most other radiation thermometers.

Optical pyrometers are useful only at temperatures at which the target is luminous, i.e., 1290°F (700°C) and above.

At one time a number of automatically adjusted optical pyrometers were available that could measure the temperatures of a surface continually. These were called "automatic brightness pyrometers". The principle of operation was the same. The function of the operator's eye was performed by a photomultiplier tube or other detector, and an electronic circuit drove the system to balance, thus determining the temperature.

These automatic brightness pyrometers have largely been displaced by less complicated radiation thermometers using semiconductor detectors that respond in the visible region.

Fiber Optics Thermometers

In fiber optics thermometers, the radiation from the target is guided to the detector by a light guide. The first such pyrometers used a 1/8-in. diameter sapphire rod to pick up energy from the target and transmit it to a thermopile detector. Contemporary fiber optics pyrometers use flexible glass fibers with or without a lens. The spectral response of these fibers extends to about 2 microns. Some are useful at target temperatures as low as 210°F (100°C).

Fiber optics thermometers are especially useful where it is difficult or impossible to obtain or maintain a clear sighting path to the target, as in pressure or vacuum chambers. Fiber optics thermometers have been used also to measure temperatures of turbine blades in gas turbines and the temperature of small objects in induction heating coils.

8-3. The Theoretical Basis for Radiation Measurements

Blackbody Radiation

Every body radiates energy to its surroundings proportional to its absolute temperature. Although the emitted radiation of a body includes all wavelengths, the region in which the amount of radiation is significant to industrial temperature measurement extends from 0.3 μm to about 20 μm. From 0.4 μm to 0.7 μm is the visible region. Radiation at wavelengths longer than 0.7 μm is in the infrared region, which humans cannot see.

The thermal energy radiated by an object is expressed relative to the energy radiated at the same temperature by a perfect radiator, traditionally called a blackbody. A blackbody absorbs all the radiation it intercepts and radiates more thermal radiation for all wavelength intervals than any other body of the same area and temperature.

Though the blackbody is an ideal, and no perfect blackbody exists, specially constructed laboratory furnaces emit radiation with an efficiency compared to a blackbody of 98% or higher. Laboratory sources with 99.98% efficiency compared to a blackbody have been constructed.

The most common approaches to realizing a blackbody are to use a spherical cavity with a small hole in the surface or a closed end tube that is much longer than its diameter. The opaque walls of the sphere or tube are held at uniform temperature. As shown in Fig. 8-2a, these constructions provide for multiple reflections of any radiation entering the opening. Thus, though the sphere or tube walls are slightly reflective, after many reflections all the energy is absorbed, i.e., at room temperature the aperture in the sphere or tube appears to be black in the visible part of the spectrum and is also nearly totally absorbing in other regions of the spectrum. At any given temperature the aperture radiates energy at nearly the same rate as a blackbody of the same size and temperature. Figure 8-2b illustrates a commercial secondary reference furnace based on a small opening in a uniformly heated spherical cavity.

Another configuration used for a blackbody source is a deep wedge, where the cavity subtends only a small angle. Multiple reflections from the sides of the wedge make it appear black. The real importance of the wedge is conceptual. Surface roughness of a product can be visualized as a multitude of small wedges. If the surface is very rough, the wedges are deep, and the product will have radiating properties that are closer to those of a blackbody than if the surface were smooth.

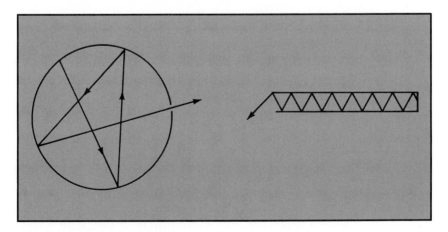

Fig. 8-2. (a) Black Body Radiation

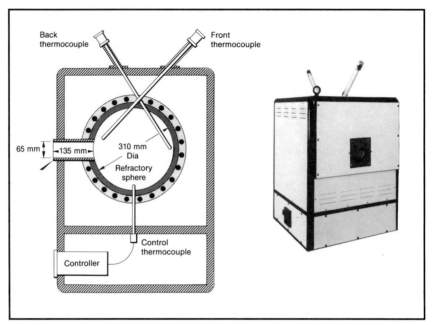

Fig. 8-2. (b) Spherical Furnace Secondary Source 200 to 1150° C
(Courtesy of Land Infrared)

Stefan-Boltzmann Law

The rate at which a blackbody radiates energy is given by the Stefan-Boltzmann Law:

$$w = \sigma T^4$$

w = watts/meter^2

σ = Stefan-Boltzmann constant, 5.6697×10^{-6}
      watts/m^2 – T^4

T = Absolute temperature, kelvins

This equation assumes that the body receiving the radiation is at absolute zero. In the practical case, the receiving body is at a temperature T_R and radiates to the blackbody at a rate $w = \sigma T_R^4$ per unit area of the receptor. Thus, the net energy reaching the receptor is:

$$w = K\sigma(T^4 - T_R^4)$$

where K is a constant taking into account the areas of the blackbody and receptor and the distance between them.

These equations give the radiation from all wavelengths in the entire spectrum. For more practical use, where the receiving object (the detector in a radiation thermometer, for example) responds significantly only to the short wavelength portion of the spectrum, the Wien-Planck and Wien Laws are more useful.

Wien-Planck Law

The Wien-Planck Law expresses the radiation emitted per unit area of a blackbody as a function of wavelength, λ, and temperature, T.

$$J_{\lambda T}\, d\lambda = C_1\, \lambda^{-5}\, (\epsilon^{C_2/\lambda T}-1)^{-1} d\lambda$$

This function is plotted for several temperatures in Figure 8-3.

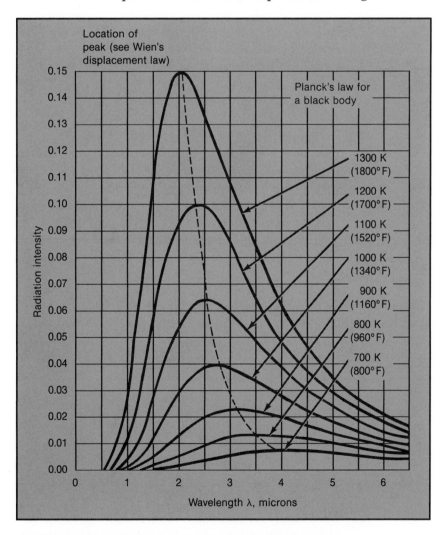

Fig. 8-3. Radiation Intensity as a Function of Wavelength and Temperature (Planck's Law)

C_1, the first radiation constant = 3.7415×10^{-16} watts/m^2

C_2, the second radiation constant = 1.43879×10^2 m·K

Wien's Law

If $\epsilon^{C_2/\lambda T}$ is much greater than 1, the Wien-Planck Law can be approximated by Wien's Law.

$$J d\lambda = C_1 \lambda^{-5} \epsilon^{-C_2/\lambda T} d\lambda$$

This expression agrees with the Wien-Planck law within 1% if λT is less than 0.003 meter ·K (3000 μm·K).

At 0.65–μm wavelength, this condition exists for temperatures below 4600K. Therefore, Wien's Law has been commonly used with high accuracy in the field of optical pyrometry.

Wien's Displacement Law

In Fig. 8-3 it can be observed that as temperature increases, not only does the amount of radiation per unit area increase, but the wavelength at which the radiation is maximum shifts to shorter wavelengths.

The value of the wavelength of maximum radiation per unit area is given by Wien's Displacement Law.

$$\lambda_m T = b$$
λ_m = the wavelength of maximum radiation, meters
T = temperature, K
b = 2.8978×10^{-3} m·K

If λ is given in μm, b = 2.8978×10^3 μm·K
Referring to Fig. 8-3,

At 2000°F (1366K) λ_m = 2.1 μm

Non-blackbody Radiation

The blackbody is an ideal concept. Real objects do not emit radiation as blackbodies, but at some lower rate. The ratio of the energy radiated by a body to that radiated by a blackbody at the same temperature is emittance (ϵ), a number less than 1.

If the body is opaque, the emittance (ϵ) is related to the reflectance (r) of the body by

$$\epsilon + r = 1$$

or

$$\epsilon = 1 - r$$

To understand that this must be true, consider an opaque object at temperature T, which is at equilibrium with its surroundings. This means that temperature T is not changing. Nearby bodies radiate energy according to their temperatures, some of which reaches the object. Some energy is reflected by the object. The remainder is absorbed. But the object is also emitting radiation proportional to its temperature, T. If the temperature is not changing, then the absorbed radiation must be balanced by the emitted radiation. If the emitted energy is greater than the absorbed energy, the object will cool. If the object absorbs more than it emits, the temperature will increase.

Thus, for any opaque body the emittance (ϵ) equals the absorbance (α).

If the object is translucent, like some plastics or glass, some of the energy incident on the object will be transmitted through the object. Therefore, in general, t is transmittance, and

$$\epsilon + r + t = 1$$

For most materials, ϵ, r, and t are functions of wavelength.

Figure 8-4 illustrates the variation in emittance with wavelength for some materials. In general, the emittance of metals is highest at short wavelengths. The emittance of ceramics is highest at long wavelengths.

There is not complete standardization of terms. The terms "emissivity" and "emittance" are often use interchangeably. Some writers use the term "emissivity" to refer to the properties of a material and use "emittance" to refer to the properties of a particular target. Thus, the emissivity of alumina might be 0.6 and the emittance of the inside of an alumina-lined oven might be 0.9 because of internal reflections.

Because emittance is wavelength-dependent and the amount of energy radiated at each wavelength depends on temperature,

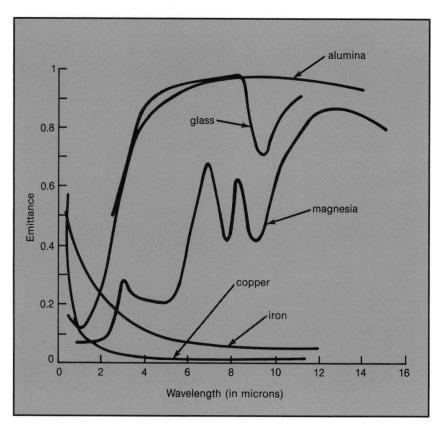

Fig. 8-4. Spectral Emittance of Some Typical Materials

the apparent emittance of a material depends on the temperature at which it was determined, at the wavelengths at which the measurement is taken.

The emittance of a metal, as determined with an optical pyrometer, which measures at a wavelength of 0.65 μm, might be 0.5. This value may not be useful if the pyrometer used to measure the temperature of the metal responds over the range of 0.3–3μm. The emittance as indicated in Fig. 8-4 is higher at short wavelengths than at long wavelengths. The proportionate amount of short wavelength energy radiated increases as temperature increases. Therefore, as shown in Table 8-1, the total emittance is lower than that measured at 0.65μm.

Table 8-1 presents ranges of both total and spectral emittances of some common materials. The value of total emittance, of course, is somewhat temperature-dependent. The ranges given are for temperatures likely to be encountered in heat treating or manufacturing operations.

	Total	Spectral = 0.65 – 1.0
Aluminum		
Unoxidized	0.02 – 0.15	0.05 – 0.25
Lightly oxidized	0.1 – 0.2	0.2 – 0.3
Heavily oxidized	0.3 – 0.4	0.4
Chromium		
Unoxidized	0.08 – 0.2	0.3 – 0.6
Lightly oxidized	0.3 – 0.5	0.5 – 0.5
Heavily oxidized	0.8	0.7 – 0.8
Copper		
Unoxidized	0.03 – 0.2	0.06 – 0.2
Lightly oxidized	0.4 – 0.5	0.4 – 0.5
Heavily oxidized	0.8	0.8
Iron		
Unoxidized	0.05 – 0.25	0.35
Lightly oxidized	0.35 – 0.5	0.45 – 0.5
Heavily oxidized	0.7 – 0.95	0.8 – 0.95
Stainless Steel		
Unoxidized	0.3 – 0.5	0.3 – 0.45
Lightly oxidized	0.5 – 0.7	0.5 – 0.7
Heavily oxidized	0.8 – 0.9	0.8 – 0.9
Gold		
Unoxidized	0.02 – 0.05	0.04 – 0.15
Heavily tarnished	0.3 – 0.45	0.6 – 0.8
Silver		
Unoxidized	0.01 – 0.03	0.02 – 0.05
Lightly oxidized	0.02 – 0.04	0.04 – 0.08
Tin		
Unoxidized	0.05 – 0.1	0.2 – 0.3
Lightly oxidized	0.25 – 0.3	0.25 – 0.45
Heavily oxidized	0.6	0.6

TABLE 8-1. Representative Emittance of Some Common Materials

For each surface condition a range of values is given for both total and spectral emittance. The lower values are for polished material. The higher values are for rough-machined material.

As materials oxidize, both the total and spectral emittances tend to increase, and the surface condition dependence of the emittance tends to decrease.

The emittances given are for materials in the open. In Unit 9, it will be shown that the effective emittance in furnaces and ovens may be higher than those given in Table 8-1.

8-4. The Response of Radiation Thermometers to Radiation

In simplest terms a radiation thermometer consists of an optical system and a detector. Both will be discussed in more detail. The output of the detector may be different at different wavelengths. The transmission of the optical elements will depend on wavelength. The output of the detector at any wavelength, therefore, is proportional to the amount radiated by the target, the amount absorbed by the optical system, and the response of the detector at that wavelength.

The calibration function of the thermometer, i.e., how the thermometer output varies with temperature, is the sum of the outputs at all wavelengths at each temperature.

This can be written, using Wien's Law, as:

$$V(T) = KC_1 \int_{\lambda_1}^{\lambda_2} \lambda^{-5} \epsilon^{-C_2/\lambda T} d\lambda$$

Any thermometer can be characterized by an effective wavelength λ_e, which is the wavelength such that

$$\lambda_e^{-5} \epsilon^{-C_2/\lambda_e T} = KC_1 \int_{\lambda_1}^{\lambda_2} \lambda^{-5} \epsilon^{-C_2/\lambda T} d\lambda$$

The effective wavelength changes with temperature, but one can calculate the effective wavelength for the calibration function of a thermometer that is applicable between any two desired temperatures.

This calculation defines the wavelength of a thermometer operating at a single wavelength that would give the same change in output between the two temperatures as the actual pyrometer does. Some manufacturers publish this effective wavelength for their pyrometers. From the effective wavelength one can estimate the N factor.

The N Factor

At a single temperature or over a narrow range of temperatures, the calibration function can be approximated in the form:

$$V(T) = KT^N$$

A pyrometer receiving radiation from a high temperature target in a wide band of wavelengths would have a calibration function approximately the form of the Stefan-Boltzmann Law, i.e., N would be near 4.

For pyrometers with restricted wavelength response, the value of N is higher.

If $V(T) = KT^N$, the pyrometer may be described as obeying the Nth power law.

It can be shown also that, approximately,

$$N = \frac{C_2}{\lambda_e T} = \frac{14388}{\lambda_e T}$$

where:

λ_e = effective wavelength in μm.
T = target temperature, kelvins

The significance of the N value of a radiation thermometer is that it allows a quick estimate of the effect of changing target emittance on the thermometer output when the target temperature is held constant.

If the target is at temperature T and the emittance is ϵ, the output is:

$$V(T) = \epsilon KT^N$$

where K is a constant that depends on the construction of the thermometer.

The output of the thermometer is directly proportional to the emittance. If emittance changes by 10% but the temperature remains constant, the output changes by 10%.

However, the output of the thermometer is interpreted in terms of temperature. How is the indicated temperature affected by a change in emittance?

If the calibration equation of the thermometer is written as:

$$V = KT^N$$

then $T = K_1 V^{1/N}$, where $K_1 = K^{1/N}$

For a change in therometer output, ΔV, the equivalent change in temperature is

$$\Delta T = \frac{K_1}{N} V^{(1/N-1)} \Delta V$$

so, $$\frac{\Delta T}{T} = \frac{K_1 V^{(1/N-1)}}{N K_1 \, V^{1/N}} \; \Delta V = \frac{1}{N} \frac{\Delta V}{V}$$

Therefore, if the output of the thermometer changes by 20%, the indicated absolute temperature will change by $\dfrac{20\%}{N}$.

Since V is proportional to ϵ, the higher the value of N, the less dependent is the temperature reading on the target emittance.

Remembering that:

$$N = \frac{C_2}{\lambda_e T}$$

it can be seen that, other things being equal, a pyrometer with the shortest possible equivalent wavelength should be chosen in order to get the highest value of N and the least dependence on target emittance changes.

The benefit of a high value of N extends to the effects of any variable that changes the output V. Thus, a dirty optical system or absorption by particles, gases, or vapors in the sighting path has less effect on indicated temperature if N has a high value.

8-5. An Overview of Radiation Thermometer Constructions

Optical Pyrometers

The functioning of the most common optical pyrometers was described in 8-2. Operation at 0.65 μm has two advantages. The first is that unoxidized metals have higher emittances at short wavelengths than at longer wavelengths. The second is that operation at 0.65 μm gives N a high value, which reduces the error caused by emittance less than 1.

At 800°C (1073 K) N is approximately 21, so the temperature of a target with an emittance of 0.8 could be measured with an emittance-caused error of approximately 1%.

Because the measurement is made in the red wavelengths of the visible spectrum, the optical components can be glass.

The schematic optical layout of an optical pyrometer is illustrated in Fig. 8-5, which shows a screen not mentioned in Section 8-2. This screen reduces the amount of radiation from very high temperature targets so that the color comparison can be made within the operating range of the tungsten strip lamp. The screen has controlled and stable absorption, so the true target temperature can be inferred from the color match.

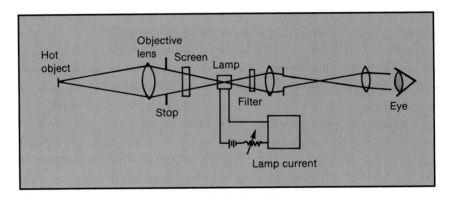

Fig. 8-5. Typical Configuration of an Optical Pyrometer

Common Radiation Thermometer Construction

Figure 8-6 illustrates the most common types of construction found in industrial radiation thermometers. The constructions in (a) and (b) are typical of instruments using detectors, such as thermopiles and silicon cells that give a stable DC millivolt output without preamplification.

The arrangement of (a) has also been used for detectors whose DC drift demands that they be used in an AC mode. A spinning disk or vibrating reed interposed between the objective lens and detector cyclically interrupts the radiation so that the detector sees pulses of radiation. The output signal of the

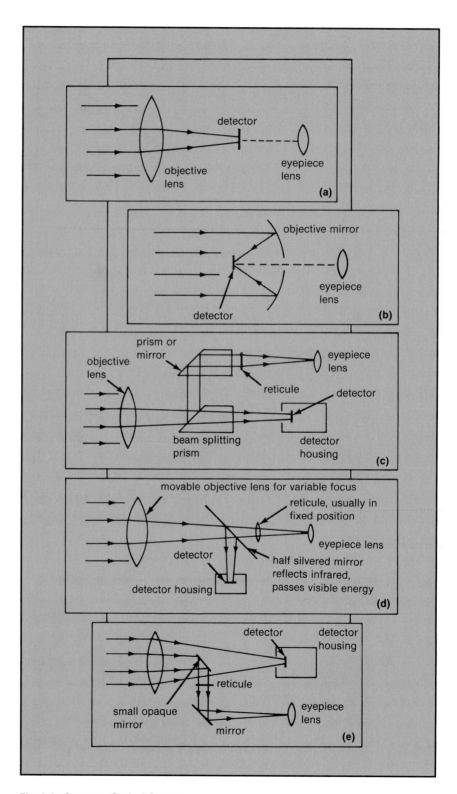

Fig. 8-6. Common Optical Systems

detector is AC. The detector package must be small enough, of course, so that it doesn't interfere with optical sighting of the target.

If the detector package is too large to permit sighting around it, the arrangements of (c), (d), or (e) are useful. Figure 8-7 illustrates a design of the type shown in Fig. 8-6(e). Optical chopping between the objective lens and the detector is common in these constructions. The back surface of the chopping disc or blade may serve as a local ambient temperature reference. The detector alternately sees the target and the modulating device, which is at local ambient temperature.

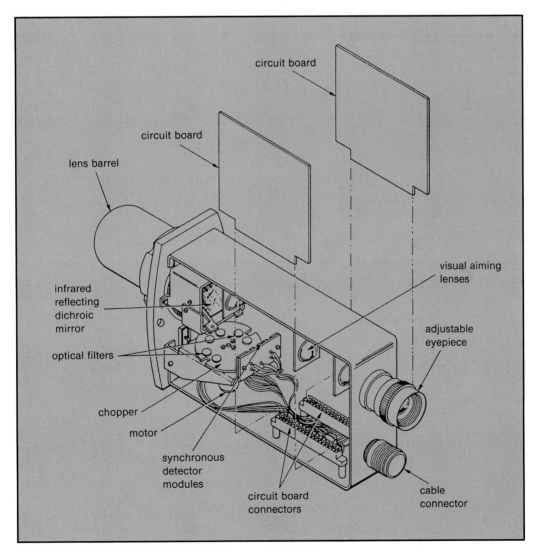

Fig. 8-7. The Arrangement of One Industrial Radiation Thermometer
(Courtesy of Williamson Corporation)

In some thermometers a local hot source, a tungsten strip lamp, or other hot surface may be maintained at a known reference temperature. The detector alternately sees the target and the reference source. The resulting AC signal can then be calibrated in terms of the unknown target temperature.

In ratio pyrometers the filters that define the pass band of the two radiation signals that are ratioed may be on the chopping disc.

Surface Temperature Pyrometers

In Section 8-3 we saw that a radiation source with characteristics close to that of a blackbody could be constructed by assuring multiple reflections within a cavity. Figure 8-8 illustrates a portable device for short-time spot measurement of the temperature of the surface of an object based on this principle.

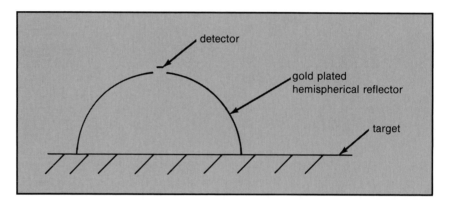

Fig. 8-8. Portable Surface Temperature Thermometer

Radiation from the target is multiply reflected from the hemispherical mirror. A detector receives radiation through a small opening in the reflector. The radiation multiply reflected between the mirror and the target appears to the detector to be from a blackbody. A commercial pyrometer using this principle can read the temperature of targets with emittance as low as 0.6 without correction.

The reflector must be placed close to the surface being measured to exclude extraneous radiation and prevent radiation losses; it can be used only for short periods because heating of the reflector will make the measurement inaccurate. Also, the energy reflected back to the target surface may cause its temperature to change.

8-6. Elements of Radiation Pyrometers

The selection of a detector and optical elements for a radiation thermometer is primarily the concern of the radiation thermometer designer, who balances out the conflicting elements of cost, accuracy, speed of response, and usable temperature range. However, as will be further discussed in Unit 9, the user should be aware of how the selection of detectors and optical elements influences the range of wavelengths over which a thermometer responds. When applying a radiation thermometer to the measurement of products in the presence of atmospheric absorption or reflections from other objects, or when trying to measure the temperature of materials such as glass or plastics, the spectral response of a pyrometer will determine whether a usable temperature measurement is possible.

Detectors

Detectors for radiation thermometers are of three types: thermal, photon, and pyroelectric.

Thermal detectors reach an equilibrium temperature proportional to the temperature of the target. The detector output is then interpreted as target temperature. The most common thermal detectors today are thermopiles and bolometers.

A thermopile consists of one or more thermocouples in series, usually arranged in a radial pattern so the hot junctions form a small circle and the cold junctions are maintained at the local ambient temperature by the mass of the supporting structure. The output, as from any thermocouple, is a DC millivoltage. Thermopiles may be manufactured from thin strips of metal, or they may be deposited on a substrate.

Bolometers were originally low mass metallic resistance thermometer elements, but today thermistor bolometers are the most likely to be used.

Usually, both thermopiles and bolometers are blackened with carbon black or other broadband absorbing material so that they respond to radiation throughout the entire spectrum, thus the common term, broadband detectors.

Because the detectors must reach a new thermal equilibrium when target temperature changes, they are relatively slow, with

time constants of a second or more. Deposited detectors may
responds in tens of milliseconds.

Photon detectors respond to incident radiation by releasing
electric charges. In lead sulfide and lead selenide photoconductive detectors, this release of charge is measured as a change in
resistance. In photovoltaic detectors, such as silicon, germanium,
and indium antimonide, the release of charge produces a voltage output.

Release of electrons and "holes" in photoconductive and photovoltaic detectors requires that arriving photons have a minimum
energy that depends on the molecular structure of the detector
material. Thus, all photon detectors have a maximum wavelength (minimum photon energy) beyond which they do not
respond. The peak response is usually at a wavelength a little
shorter than the cutoff wavelength.

The approximate peak response wavelengths of some photon
detectors are:

Material	Wavelength, μm
Silicon	0.9
Germanium	1.7
Lead sulfide	2.5
Indium arsenide	3
Indium antimonide	6

Many modern radiation thermometers use photon detectors
instead of thermal detectors despite the limited range of wavelengths over which they respond. This limited response may
help tailor the thermometer response to suit the needs of a
particular application. But, even in general-purpose thermometers, a photon detector may be preferred. The loss in output
because of the limited spectral response is more than offset
by the fact that within the range of useful wavelengths the
sensitivity of photon detectors is 10^3 to 10^5 times that of a
thermopile.

Pyroelectric detectors respond to changes in received radiation
with a changing surface charge. Because it responds to changes
in incoming radiation, the detector need not reach thermal
equilibrium when the target temperature changes. However,
the incoming radiation must be chopped, and the detector signal cannot be utilized directly. The detector changes can be

likened to a change in charge of a capacitor, which must be read with high impedance circuitry.

Pyroelectric detectors bear a radiation-absorbent coating, so they can have a spectral response as broad as thermal detectors, or, by selecting the characteristics of the coating, a more restricted response.

Optical Systems

The optical system of a radiation thermometer may be composed of lenses or mirrors or combinations of both. In general, the reflectivity of mirrors is not significantly dependent on wavelength over the range of wavelengths used for industrial temperature measurement. Therefore, mirror systems do not determine the spectral response of the pyrometer. A mirror system must usually be protected from dirt and physical damage by a window. The characteristics of the window, and those of the detector, determine the band of wavelengths over which the thermometer responds.

Mirror systems have been common mainly in fixed focus optical systems. To vary the focus of an optical system requires that at least one element be movable. Providing for this motion in mirror systems is often more complicated than in lens systems. The selection of lens and window materials is always a compromise between the optical and physical properties of the material and the desired wavelength response of the pyrometer.

The properties of some typical lens and window materials are given in Table 8-2.

8-7. Field of View

The field of view of a radiation thermometer is a statement of the size of the target at a specified distance from the thermometer. This may be stated in the form of a diagram, a table of target sizes versus distance, sometimes simply as the target size at the focal plane and the distance to the focal plane, or as an angular field of view.

If the output of a radiation thermometer is to be the same as its calibrated output, the target must fill the field of view. If the target fills the field of view at any distance from the thermometer, the output will be the calibrated output, or nearly so, as discussed below.

Material	Refractive Index	Cut-off Wavelength – μM	Mechanical & Chemical Stability	Remarks
Glass	1.5	2.8	Excellent	Excellent transmission in the visible
Fused Silica	1.4	4	Excellent	Excellent transmission in the visible
Calcium fluoride	1.4	10	Good	Good visible transmission Relatively high chromatic aberration
Arsenic trisulfide	2.35	12	Good	Low transmission in visible. Low chromatic aberration
Zinc Sulfide	2.25	14	Good	Very low transmission in visible
Polyethylene	—	20	Poor	Sometimes used as window. Easily distorted or heat crazed.

TABLE 8-2. Properties of Lens and Window Materials

Figure 8-9 illustrates the field of view of a perfect radiation thermometer.

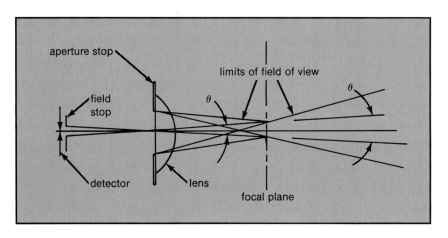

Fig. 8-9. The Field of View of a Perfect Radiation Thermometer

The image of the field stop in the focal plane in most thermometers is larger than the diameter of the field stop. θ is the angular field of view.

Between the aperture stop and the focal plane, the field of view is determined by the stop diameter and the image diameter.

Lines drawn from the extremities of the image to the extremities of the aperture stop enclose the field of view.

Beyond the focal plane the field of view is determined by rays extending from the extremities of the aperture stop through the extremities of the image in the focal plane.

Figure 8-9 shows that only if the image in the focal plane is large relative to the lens, and the focal plane is many lens diameters from the lens, will the actual field of view be given by the angle θ. The actual field of view is always larger.

Most thermometers have images in the focal plane that are the same diameter as the lens or smaller, so the statement of an angular field of view is not sufficient definition. Therefore, diagrams or tables are usually used to define the field of view.

In practice, any statement of the field of view is only an approximation because of spherical and chromatic aberration.

Spherical aberration is caused by the fact that rays hitting the lens remote from its axis are bent more than rays passing the lens near its axis. A circular field stop is therefore imaged as a circle with a fuzzy halo around it. Mirrors also have spherical aberration.

Chromatic aberration occurs because the refractive index of optical materials changes with wavelength. At shorter wavelengths the refractive index is lower, and rays exiting are bent more and focus nearer the lens. Rays of longer wavelengths are focused farther from the lens. The image of a field stop over a band of wavelengths is therefore a fuzzy image.

Other and generally smaller contributions to fuzziness of the field of view are scattering of rays of light by imperfections in the optical material and reflections from the internal parts of the radiation thermometer. Good quality materials minimize the former cause. Roughening and blackening of interior surfaces to reduce reflections reduces the latter effect.

There are no standards at present for stating the field of view of radiation thermometers. Some manufacturers state a field of view that includes the effects of aberrations; others do not.

If the target size and stated field of view are nearly the same, it may be wise to determine the field of view experimentally.

Sight the thermometer on a target that gives a steady, uniform source of radiation. At the focal plane, interpose a series of apertures of different diameter between the thermometer and the target, and plot the output of the pyrometer against the aperture area. For aperture areas less than the nominal target area, the output of the thermometer should increase directly proportional to aperture area. Above the nominal target area the output should increase only a small amount with further increases in aperture area. A perfect thermometer would show no increase, but none such exists. Increases of a few tenths of a percent in output for each doubling of the aperture area indicates that the nominal field of view takes into account the effects of aberrations. If these are not taken into account, the thermometer may show a significant increase in output as the viewable target area is increased above the nominal value.

8-8. Signal Conditioning

Linearization

The calibration curves of detector output versus temperature of all detectors is nonlinear because, as was explained in Section 8-3, the equations relating the amount of radiation emitted by an object are inherently power functions. If the output of a detector is approximated by $\epsilon = KT^N$, N has a value of at least 4, and it can be as high as 20.

Before the advent of inexpensive microprocessors, the advantage of high N values in reducing the effects of varying emittance was offset by the fact that, if the output temperature varies as a high power of temperature, the useful range of temperature measurement with an indicator or recorder of fixed span is very low.

If N = 15, for example, and a recorder reads 100% of scale at 1000°C, the recorder would read approximately 820°C at 10% of scale. If the target temperature were expected to be outside this relatively narrow range, some form of linearization or some form of automatic range switching is necessary.

Inexpensive microprocessors now permit such signals to be linearized easily. Many radiation thermometers now provide a linear output as a standard feature.

Sample and Hold

The sample and hold function is useful where it is desired to measure a product temperature whenever a selected event triggers the measurement. The thermometer measures temperature at that instant, disregarding earlier or later measurements. Analog sample and hold circuits always exhibited a slow drift of the measurement during the "hold" period, but modern digital instrumentation can hold the value without degradation for an indefinite time.

Peak and Valley Picking

In some applications the temperature of interest is the highest temperature within the field of view during a given period. This might be true of a target that is occasionally obscured by smoke or cannot be viewed continuously (as a series of objects, such as bottles). The electronics interface then can be programmed to "remember" the highest temperature it saw in the sampling period.

Valley picking is the inverse of peak picking, useful where the lowest temperature measured is the preferred value for control.

Averaging

Averaging is usually used to prevent rapid excursions of the temperature reading from the average value from being recorded or causing noise in the control system. A common way to average is to slow down the response of the electronics package with a low-pass resistance-capacitance (RC) filter; but the equivalent function is often now performed in software in microprocessor-based signal conditioners.

Exercises

8-1.   *When is radiation thermometry likely to be the temperature measurement method of choice?*

8-2.   *What are some of the disadvantages of radiation thermometry?*

8-3.   *What is a total radiation thermometer?*

8-4.   *How does a ratio thermometer work, and when is it useful?*

8-5.   *What is the principle of operation of an optical pyrometer?*

8-6.   *What is the principle of a blackbody? How can one be constructed?*

8-7.   *Why is radiation thermometry often said (erroneously) to be based on the fourth power law?*

8-8.   *Why is Wien's Law useful?*

8-9.   *What is the significance of the Wien Displacement Law?*

8-10. *Describe emittance.*

8-11. *How is the absorbance, α, related to the emittance, ϵ, of an opaque body?*

8-12. *Why is the total emittance of a metal higher at higher temperatures?*

8-13. *What is the effective wavelength of a thermometer? What is it useful for?*

8-14. *Why is a high N factor always desirable?*

8-15. *How does one obtain a high N factor?*

8-16. *What are the common thermal detectors? What are their advantages? What are their disadvantages?*

8-17. *How is the response of photon detectors different from those of thermal detectors?*

8-18. *Why is glass not used in all radiation thermometer optical systems?*

8-19. *Why is a field of view specification always, to some extent, an approximation?*

Unit 9:
Selection, Application, and Calibration of Radiation Thermometers

UNIT 9

Selection, Application, and Calibration of Radiation Thermometers

Unit 9 builds upon the facts about the theory and construction of radiation thermometers presented in Unit 8 to examine how they affect the selection and application of radiation thermometers.

Learning Objectives — When you have completed this unit you should:

A. Know how to select a radiation thermometer for a specific temperature measurement.

B. Know how to reduce instrument error.

C. Know how to maintain and calibrate radiation thermometers.

9-1. Criteria for Selecting a Radiation Thermometer

Requirements of the Application

Any prospective user of a radiation thermometer is driven by a desire for low initial investment and low maintenance cost, accuracy consistent with the needs of the process, and high reliability. These qualities are desired for any instrumentation, but the selection of the right radiation thermometer is often driven almost entirely by the requirements of the measurement.

One can narrow the range of choice by listing first the essential needs of the measurement. For first level sorting, consider speed of response, target size, and target temperature.

Dozens of radiation pyrometer models are available. Each represents a different compromise in the choice of target size, speed of response, and usable temperature range to meet a variety of application needs. In addition, the standard models can often be altered to meet special demands, such as trading off a higher target temperature for a smaller target size. The observations given below should be viewed only as typical, not definitive limits of choice. However, when radiation thermometry is being considered as an alternative to some other measurement strategy, these factors provide a basis for judging whether it is likely that radiation pyrometry is practical and whether it is economically justified.

Always, it is wise to take advantage of the experienced counsel available from radiation thermometer manufacturers.

Speed of Response

With few exceptions, if response to 90% of a step change in less than a few seconds is required, pyrometers using thermal detectors will not be suitable. A pyrometer using a photon detector will be necessary, though it may be a relatively simple, low-cost model.

Target Size

Thermometers with target sizes of 0.3 to 1 inch diameter in a focal plane lying 1.5 to 3 feet from the lens are most common. If a target size in this range is needed only to sight on a large target through a small opening in a furnace, a simple pyrometer whose target size increases rapidly with distance beyond the focal plane may be used. Otherwise, a thermometer with more sophisticated optics and signal conditioning may be required.

Target Temperature

If the temperature to be measured is below about 750°F (400°C), one of the simpler designs of radiation pyrometers is unlikely to be satisfactory. A more sophisticated thermometer using optical chopping will probably be required.

For most applications, speed of response, target size, and target temperature are sufficient to pick the models that can be seriously considered and to allow a first judgment on whether radiation thermometry is an economically viable alternative to other methods. Section 9-2 discusses some common problems of radiation thermometry and common approaches to resolving them. In some cases, these problems will further constrain the kind of thermometer that can be used.

9-2. Application of Radiation Thermometers

Emittance and Reflectance Corrections

In the early days of radiation thermometry, an uncalibrated emittance adjustment in the recorder or indicator that received the signal from the thermometer changed the span of the instrument to compensate for a target emittance less than 1. The user determined the temperature of the target by a temporarily installed thermocouple, a portable optical pyrometer, or from experienced assessment of product quality and then adjusted the instrument to read the correct temperature.

Many instruments now have calibrated emittance adjustments. The adjustment may be set at a value determined from tables of emittance found in the manufacturer's literature or in handbooks. The values in such tables are, of course, typical values. For highest accuracy, an independent determination of temperature or emittance is necessary. If the investment is justified, the emittance of a representative sample can be determined in the laboratory at the wavelength at which the thermometer measures. It is not necessary to measure at expected target temperature unless there is reason to believe that the surface properties of the material change with temperature because of oxidation or change of state of a coating.

If the response of the thermometer is broadband, the effective emittance may change with temperature because the predominant contribution of energy shifts to shorter wavelengths as temperature increases. If this is suspected, the emittance determination should be made at the expected temperature of the target.

The emittance values listed in most tables have been determined with a pyrometer sighted perpendicular to the target. If the actual sighting angle is more than 30° to 40° from the normal to the target, the emittance values may not be applicable. An auxiliary measurement of emittance or temperature is needed.

In many applications, such as vacuum heat treatment, the radiation thermometer sights through a window. The emittance correction must include provision for energy lost by reflection from the two surfaces of the window as well as absorption in the window. For glass, about 4% is reflected from each surface in the visible and near infrared ranges, so the effective transmittance is 0.92. The reflection loss from windows of other material can be calculated from the index of refraction of the material at the wavelength of measurement.

For each surface, the reflectance, $r(\lambda)$, is given by:

$$r(\lambda) = \left(\frac{\eta - 1}{\eta + 1} \right)^2$$

where η is the index of refraction at the wavelength λ.

The absorption loss can usually be neglected.

Target and Surroundings at Different Temperatures

For an opaque object, the measured temperature is given by:

$$T_m = \epsilon_\lambda T_B + (1 - \epsilon_\lambda) T_S$$

where:

T_m = Measured (apparent) temperature of the object
ϵ_λ = Spectral emittance of target at wavelength λ
T_B = Actual target temperature
T_S = Temperature of surroundings (assuming that the surroundings are at a reasonably uniform temperature)

The $(1 - \epsilon_\lambda)$ term represents the reflectance of the target at the wavelength λ.

If the surroundings are not uniform or if a hot source only partially fills the field of view of the thermometer, the second term must be multiplied by a constant less than one. This circumstance requires special investigation. It would be better to try to shield the target so that the hot objects are not seen in the field of view of the thermometer (see below).

Target and Surround at Nearly the Same Temperature

From the equation given above it can be seen that if the surround temperature, T_S, is the same as the target temperature, the indicated target temperature will be accurate. If one can assume that ϵ_λ and T_S are reasonably constant, one can estimate the error for conditions when T_S is not equal to T_B. Many manufacturers can assist in estimating this error, which depends on the thermometer's spectral response.

In many applications (batch heat treating, for example) at the end of a long soak period to stabilize metallurgical properties of the work, the work temperature and the furnace temperature are nearly equal, so the effective emittance of the target is close to 1.

In other applications, as in a molten salt bath, the radiation thermometer is sighted into a closed-end tube whose temperature is the same as the material surrounding it. The thermometer is sighting into a cavity whose length is much greater than its diameter, and, therefore, it is effectively measuring black-body temperature.

Target at Higher Temperature than the Surroundings

If the target is hotter than the surroundings, the second term in the equation can usually be neglected. Many industrial targets have emittance greater than 0.5. Because T_S is lower than T_B, its radiated energy is much lower, and only a portion $(1-\epsilon_\lambda)$ is reflected into the thermometer field of view. The thermometer, therefore, will read a temperature close to $\epsilon_\lambda T_B$.

To minimize the emittance error and also to minimize the effect of the second term, it is desirable to use a pyrometer with a high N value.

If the target emittance is lower than 0.5, as it is for targets of aluminum, galvanized steel, or tin-coated steel, the reflected term may not be insignificant. For such applications the guidance of someone experienced in such measurements should be sought.

Target at Lower Temperature than the Surroundings

In many applications the target is being heated in a furnace whose walls are at a much higher temperature than the target. In others the target is being heated by high temperature electrical or gas-fired heating elements. The second term in the equation cannot be neglected. It may even be larger than the first term. Two approaches to minimizing the contribution of the surround can be used.

The first method is possible if the target is fixed in position, is flat, and reflects specularly, i.e., like a mirror, rather than diffusely like a matte surface. This is the case with flat glass. The method is to arrange the sighting path of the thermometer so that it sights perpendicular to the target surface and so that the reflected field of view intercepts the cool thermometer itself (see Fig. 9-1).

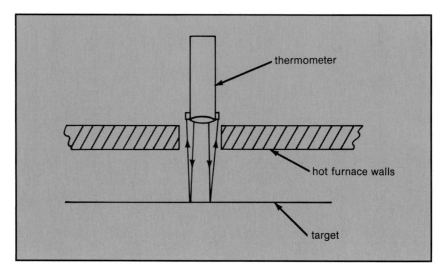

Fig. 9-1. Sighting on a Specularly Reflective Target in Hot Surroundings

Most targets are not specular reflectors. Their roughened or oxidized surfaces reflect diffusely; i.e., a ray striking them is reflected in many directions. To measure the temperature of such targets in the presence of hotter surroundings, it is necessary to shield the field of view of the thermometer so that energy from hot objects does not enter.

One approach is to sight the thermometer through an open-end sighting tube as illustrated in Fig. 9-2 (a). If the sighting tube is close to the target, less of the energy from hot sources can be seen by the thermometer. This method is ineffective if the sighting tube itself becomes hotter than the target.

Where a sighting tube won't work, a cooled shield may be used, as in Fig. 9-2 (b). The shield must be water cooled so that it does not reach the temperature of the furnace walls. The shield must be large enough so that D/H is 2 to 4. This method cannot be used for slowly moving or stationary targets because the cooling effect of the shield would change the target temperature.

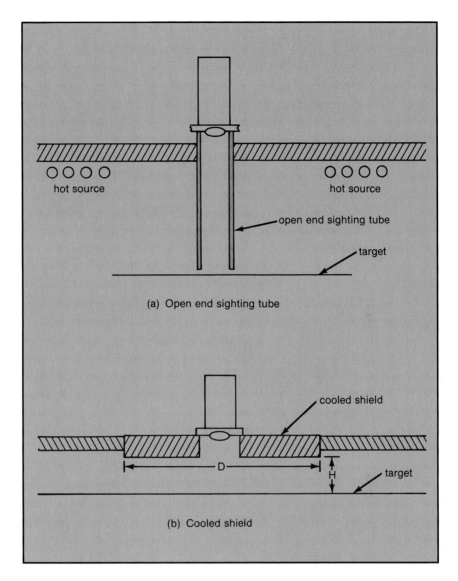

Fig. 9-2. Shielding To Reduce the Effect of Hot Sources

An uncooled shield may be used when it is blocking out radiation from small, high temperature sources that will not heat it significantly.

Absorption in the Sighting Path

A radiation thermometer infers temperature from the radiation it receives. Any radiation absorbed by gases or particles in the sighting path will lower the indicated target temperature. Conversely, any energy radiated by hot gases or particles will raise the indicated target temperature.

The influence of absorbing materials (water vapor and carbon dioxide being the most common) can be reduced by judicious selection of the wavelengths at which the thermometer responds. The influence of hot particles can be eliminated only by ensuring that they do not enter the sighting path or, if they are only transiently present, by peak or valley picking, so that the thermometer responds only to the readings taken when the particles are not present.

In many cases an open-end sighting tube, purged with a low velocity stream of clean air, mounted so the open end is near the target will provide a clear sighting path, free of interfering particles or gases, without affecting target temperature.

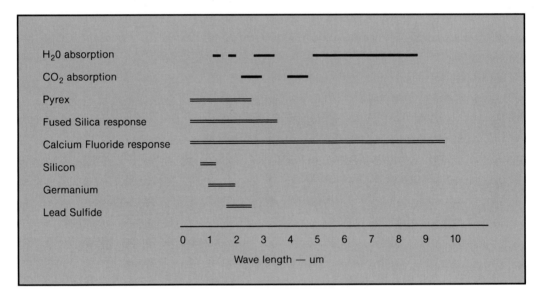

Fig. 9-3. Absorption bands of CO_2 and H_2O and response bands of some pyrometers

Where interference is caused by water vapor or carbon dioxide, a thermometer that operates in regions of the spectrum where these gases do not absorb strongly can be selected. Figure 9-3 illustrates the regions of significant absorption by these two materials. The figure is indicative only. The amount of energy absorbed depends on the length of the sighting path and the concentrations of CO_2 and water vapor. But it can be seen that a thermometer with a Pyrex lens transmitting to 2.8 μm will be less affected by CO_2 and water vapor than one with a fused silica lens that transmits to 3.8 μm. A pyrometer with a silicon detector operates outside the absorption bands of both H_2O and CO_2. The absorption error is nil.

Transparent Targets

The temperature of "transparent" targets such as glass or plastic films must usually be measured at a wavelength where the transmission is low so that hot objects behind the target do not interfere with the measurement. Preferably, one would choose a wavelength at which transmission is nearly zero.

$$T = (1-r_\lambda)^2 \, \epsilon^{-\alpha x}$$

where:

T = Transmission
r_λ = Reflection coefficient calculated from the index of refraction
x = Thickness of film
α = Spectral absorption coefficient at the wavelength of interest

Most glasses are opaque at wavelengths above about 5 μm if they are 3 mm or more thick, so that a band-pass pyrometer operating from 5 to 8 μm could be used. From 8 to 10 μm, the emittance of the glass decreases because of its high reflection; so measuring beyond 8 μm is not desirable.

If the sighting path is long, the measurement will be affected by the water vapor absorption band. For a 48-inch sighting path, water vapor absorbs about 10% of the energy in the 5.5 to 7.5 μm band. Therefore, a narrow band thermometer centered at about 5 μm is the preferred device to use in the presence of water vapor.

Thin-film plastics of the polyethylene type are transparent over wide regions of the spectrum, but they are opaque at 3.43 μm. Polyester-type films are opaque at 7.9 μm. A thermometer incorporating a narrow band filter centered on one of these wavelengths can be used to measure film temperature. Little energy is available because of the narrow pass band and relatively low target temperature, so chopped optics are a necessity. The thermometer, therefore, is likely to be more costly than those used for other applications.

9-3. Radiation Thermometer Accessories

Every radiation thermometer is given an ambient temperature rating by its manufacturer, above which accuracy and reliability diminish. When sighting on objects in high temperature furnaces and ovens, it is often necessary to mount the thermometer on or very close to the furnace. To maintain the temperature of the thermometer within its rated range, manufacturers provide a variety of air- or water-cooling means. Some provide for cooling the thermometer enclosure directly. Others provide accessory fittings that can be mounted between the pyrometer and the furnace to reduce the amount of heat conducted from the furnace. These fittings may also provide means for holding sighting tubes. Fused shutter assemblies are also available, which will prevent radiant energy from reaching the pyrometer housing if its rated temperature is exceeded. Figure 9-4 illustrates some typical accessories.

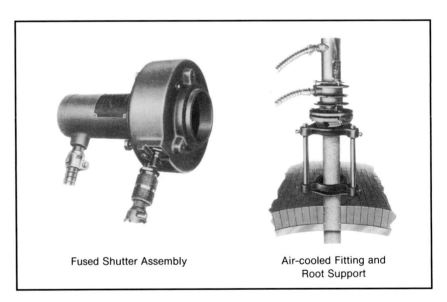

Fused Shutter Assembly   Air-cooled Fitting and
                                          Root Support

Fig. 9-4. Typical Accessories for Radiation Thermometer (Courtesy Honeywell Inc.)

9-4. Calibration of Radiation Thermometers

Optical Pyrometers

Secondary standard tungsten strip lamps are available with National Bureau of Standards calibration at 0.65 μm. These are used primarily for the calibration of optical pyrometers.

Radiation Thermometers

Two methods of calibrating radiation thermometers are common. One method is to use a high quality laboratory blackbody furnace whose temperature can be controlled accurately to provide a source of known characteristics. The second method is to use a transfer standard, a pyrometer whose calibration is known to be accurate. The calibration furnace need not be a blackbody, but need only approximate blackness. Its emittance may be less than 1.0 as long as its temperature is uniform and the emittance is the same over all wavelengths of interest, i.e., the source is a graybody. Its temperature need only be known to a moderate degree of accuracy.

In practice, the thermometer to be calibrated and the reference standard thermometer are each sighted at the graybody source. The working thermometer is adjusted to agree with the reading of the reference standard.

For most industrial users of radiation thermometers, this method is preferred, not only because a less costly furnace can be used, but also because the method can eliminate several uncertainties of calibration.

In Unit 8 it was shown that the field of view of radiation pyrometers is always somewhat fuzzy because of chromatic and spherical aberrations.

If a thermometer with a nominal field of view of 1-3/4-in. at 24-in. is sighted at a 2-in. target from a distance of 24-in. and then at several smaller distances, some increase in output will usually be observed as the distance from the furnace is decreased. The size of this effect may change as target temperature changes.

If the reference standard and the working thermometer both sight on the same target from the same distance, this effect is unimportant. The working thermometer need only be adjusted to yield the same reading as the reference thermometer.

Similarly, if a particular thermometer operates in a band where atmospheric water vapor absorption is significant, this comparison with a reference standard under identical conditions eliminates inaccuracies caused by absorption during the calibration procedure.

Exercises

9-1. What three factors should be considered first when evaluating the possible use of a radiation thermometer for a particular application?

9-2. Is it necessary to know the emittance of the target accurately?

9-3. If the emittance of a sample is measured in the laboratory, when is it necessary to measure it at the normal target temperature?

9-4. How does one correct for energy lost in a viewing window?

9-5. Why is the emittance of a billet in a reheat furnace or a coil of steel in an annealing oven close to 1?

9-6. What is a specular reflector?

9-7. When must the shield used to block out radiation from the surroundings be cooled?

9-8. How can one deal with energy losses caused by water vapor or carbon dioxide absorption?

9-9. How can one deal with errors caused by hot particles in the sighting path?

9-10. Can transparent or translucent targets be measured?

9-11. Why is the comparison method of calibrating radiation thermometers used in industrial thermometry?

Unit 10:
Some Other Methods of
Measuring Temperature

UNIT 10

Some Other Methods of Measuring Temperature

This unit discusses temperature sensors that may be useful in special circumstances, though they would not usually be used for routine measurements in the plant.

Learning Objective — When you have completed this unit you should:

A. Be familiar with a variety of laboratory temperature measuring devices.

10-1. Glass Stem Thermometers

Although most of us think of glass stem thermometers as used clinically to determine body temperature or around the house to measure ambient air temperature, the useful range for glass stem thermometers extends from -200 to 600°F (-100°C to 350°C).

Glass stem thermometers are inexpensive, but they are suitable only for local measurement. They are often difficult to read and are easily broken.

The typical glass stem thermometer consists of a glass reservoir containing mercury, alcohol, pentane, or some other liquid that expands as its temperature increases and the stem with a small diameter bore into which the liquid expands. The bore is usually evacuated above the liquid level, but it may be filled with inert gas to increase the usable temperature range.

If the capillary is constricted just above the reservoir, the thermometer will read the highest temperature to which it is exposed, as in the common clinical thermometer. The constriction prevents the cooled fill liquid from entering the reservoir again when the bulb temperature decreases from the maximum it has sensed.

10-2. Bimetallic Thermometers

Bimetallic thermometers can be used over about the same range as glass stem thermometers. They function because most metals expand as their temperature increases, but not all metals expand by the same amount. If two strips of different metals are bonded together and heated, the composite strip will bend towards the side of the metal with the lower expansion rate. Usually, one of the strips is a nickel alloy with very low coefficient of expansion.

If a bonded element is wound into a spiral or helix, one end of which is fixed, the motion of the free end can be calibrated in terms of temperature. Many domestic meat thermometers and ambient temperature indicators use bimetallic elements.

In industry the elements can be made to indicate on dials 2 to 5 in. in diameter, with stem lengths up to two feet.

Bimetallic thermometers are more rugged than glass stem thermometers but less accurate. Like glass stem thermometers, they are suitable only for local indication.

10-3. Pyrometric Indicators

Pyrometric indicators are devices that indicate the temperature to which they have been exposed by exhibiting a change in shape, dimension, hardness, or color. Most are suitable for one use only, though some materials that change color are reusable.

Ceramic Indicators

Some ceramic indicators are conical or rod-shaped. Upon exposure to heat, the cone tip or the rod sags. The temperature at which sagging of the cone tip occurs depends on the material. By using a series of cones, temperature differences of a few degrees may be estimated.

The amount of sag of the rods, which are supported at the ends, is also interpreted as temperature.

Other indicators depend on measuring the diameter of a hole or a controlled dimension to determine the temperature to which the indicator has been exposed.

Because the deformation of ceramic indicators is a time-temperature phenomenon, these devices do not give an exact temperature measurement.

Ceramic indicators are useful over the range 1100° to 3700°F (600° to 2000°C).

Color Indicators

Color indicators change color at a predetermined temperature. Most are irreversible, but some can be used repeatedly because the color change is reversible.

Irreversible indicators are available in the form of labels, usable from 100° to 2500°F (38° to 1371°C). A dot on the label turns black at the temperature rating. Accuracy is ± 1% of rating. Labels are available with multiple dots whose transition temperatures are spaced at 10°, 25°, or 50°F (−12°, −4°, or 10°C) increments.

Melting Point Indicators

Melting point indicators change state at the rated temperature. When applied with a crayon or as a lacquer, the mark is dry and chalky. At the rated transition temperature the material liquefies and becomes glossy. The change in state is not reversible. Range of application is the same as color indicators. Accuracy is ± 1% of rating.

These materials are also available in small pellets for applications where heating is slow and melting can be observed.

Melting point standards are also available in liquid or granular form for use from 125°F to 600°F (52°C to 316°C). These are calibrated to ± 1°F of the nominal transition temperature.

Application of Pyrometric Indicators

The applications of pyrometric indicators are of three general types. One type of application is in determining the temperature of an object of such low mass that even the smallest thermocouple would conduct heat away from the object and change its temperature. One example is a very thin sheet of shiny metal, whose temperature cannot be accurately measured by a radiation thermometer unless the pyrometer is calibrated at known temperatures. By using a melting point indicator or color change indicator, the radiation thermometer can be calibrated at several temperatures of the sheet.

Another kind of application is where the process is to be controlled by an indirect temperature measurement, i.e., by measuring the air temperature in a furnace, but it is desired to ensure that the temperature at which the furnace is controlled will bring the work to the desired temperature. Typical of this kind of application is one where objects on a belt are carried through a furnace whose temperature is controlled at a constant value. The belt speed and furnace temperature must be adjusted to ensure that the desired product temperature is reached.

A thermocouple traveling with the product is oftentimes not practical. Therefore, color change indicators are used.

A third application, now quite common, is to affix color change labels to a product so that the manufacturer can determine whether the user has subjected the product to a temperature higher than its rated ambient operating temperature should it be returned for failing to perform.

10-4. Thermistors

Thermistors are temperature-sensitive resistance elements made from oxides of nickel, manganese, copper, titanium, and other metals that are sintered at high temperatures. The temperature resistance characteristic of thermistors is exponential. Most thermistors have a negative temperature coefficient, but some are available with a positive temperature coefficient. A typical thermistor with a resistance of 1200 ohms at 40°C will have a 120-ohm resistance at 110°C. This thermistor will decrease in resistance by a factor of about 2 for every 20°C increase in temperature. Thus, the thermistor can be used to measure very narrow spans.

The lack of interchangeability of thermistors was once one of the strongest factors against their widespread use, but thermistors are available now that will match the manufacturer's calibration curves to ± 0.1°C up to a maximum operating temperature of 212°F (100°C).

Other thermistors are available for use up to temperatures of 572°F (300°C).

Because of their nonlinear temperature-resistance characteristics, thermistors have customarily been used for measurements over narrow ranges. However, by using resistors with very low temperature coefficients in series and parallel with one or more thermistors, probes can be constructed whose output is linear with temperature. The sensitivity of the measurement is, of course, reduced. One manufacturer supplies probes that are linear to $\pm 0.065 °C$ over the range $-22°F$ to $212°F$ ($-30°$ to $100°C$) to and interchangeable to $\pm 0.15 °C$. When the linearized probe is fed from a constant voltage source, the voltage output may be 400 times that of a Type J thermocouple, i.e., 20 mv/$°C$ compared to 50 μv/$°C$.

Thermistors are available in a wide variety of styles, such as tiny beads, discs, washers, and rods. They may be encased in glass or plastic, or they may be bare, as required by the application.

The resistance of thermistors may be measured using a Wheatstone bridge, by driving the thermistor from a constant current source and measuring the voltage across it or by driving it with a constant voltage and measuring the current through it.

Because the resistance of the thermistor can be chosen to be thousands of ohms over the measured temperature range, lead resistance errors can usually be neglected. In principle, when measuring the voltage drop across the thermistor, the voltage-measuring device should have an input resistance about 1000 times the highest thermistor resistance in the operating range. Modern semiconductor circuitry, however, can easily meet this requirement.

Many users of thermistor measurements purchase complete measuring packages that consist of precalibrated probes and indicating or controlling devices.

Although thermistors are especially useful for laboratory measurements, they have not found wide use in industrial temperature measurement, primarily because their use is limited to relatively low temperatures and there are no industry-wide, standardized resistance-temperature characteristics.

10-5. Semiconductor Sensors

A number of laboratory applications of silicon diodes and transistors for temperature measurement have been described in the literature over the years.

Diodes are energized with a constant current, typically 500 μa or less. The voltage across the diode is linearly dependent on absolute temperature. Sensitivity can be tens of millivolts per °C.

Transistors are generally operated with a constant emitter current. The base bias to maintain this constant current is linearly dependent on temperature.

In both cases, the sensor must be individually calibrated.

Some systems using this technique have been offered for sale from time to time but must have been custom built for specific laboratory applications. The range of temperature measurement is to about 300°F (150°C).

This technique has also been used for automatic cold junction compensation of digital meters reading thermocouple voltages.

10-6. Heat Balance Thermometry

The measurement of the surface temperatures of electronic components and other small objects is difficult. Very small thermocouples can sometimes be used, but they must be attached with great care with thermally conducting cement to ensure that the thermocouple temperature represents the actual surface temperature. The measurement is always of doubtful accuracy, because some heat flows from the component through the sensor and its leads and the temperature drop between the surface and the sensor caused by this heat flow is unknown.

One approach to such measurements, which has been marketed, is a probe with two temperature sensors, as shown schematically in Fig. 10-1. In the absence of a heat source, heat flows from the surface at temperature T_s and causes a temperature gradient along the probe such that T_2 is less than T_1. If heat is supplied to the probe so that $T_1 = T_2$, there is no heat flowing axially in the probe, and $T_s = T_1 = T_2$.

Since the system must adjust the input to the heater until $T_1 = T_2$, the measurement is relatively slow.

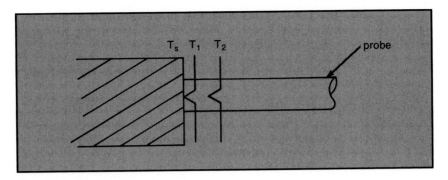

Fig. 10-1. Heat Balance Probe

Appendix A
Suggested Reading
and Study Materials

APPENDIX A

Suggested Reading and Study Materials

Benedict, Robert P., *Fundamentals of Temperature, Pressure and Flow Measurement*, Wiley, 1977.

"Manual on the Use of Thermocouples in Temperature Measurement," STP470, 3rd ed., American Society for Testing and Materials, 1981.

Kerlin, Thomas W., and Robert L. Shepard, *Industrial Temperature Measurement*, Instrument Society of America, 1982.

Considine, Douglas M., *Process Instruments and Controls Handbook*, 3rd ed., McGraw-Hill, 1985.

Liptak, Bela G., *Instrument Engineers' Handbook*, vol. 1., Chilton, 1969.

In addition to these books, many vendors of measuring elements and associated apparatus provide valuable application notes in their catalogs and in technical monographs.

Appendix B
Solutions to All
Exercises

APPENDIX B

Solutions to All Exercises

UNIT 2

2-1.   (a) $953°C = 953 \times \dfrac{9}{5} + 32 = 1747.4°F$

      (b) $1000°F = (1000-32)\ \dfrac{5}{9} = 537.78°C$

                     $537.78°C = 537.78 + 273.16 = 810.94\ K$

      (c) $1800\ K = (1800-273.16)°C = 1526.84°C = 2780.31°F$

      (d) $1800\ K = 1526.84°C = (2748.31 + 32)°F$
                       $= 1526.84°C = 2780.31°F$
                       $= 1526.84°C = (2780.31 + 459.69)°R$
                       $= 1526.84°C = 3240°R$

            or alternatively: $1800\ K = \dfrac{9}{5} \times 1800 = 3240°R$

2-2.   The temperature of the water-ice bath is an excellent approximation to the triple point of water. In highly precise measurements the bath is enclosed so that the vapor above the water-ice mixture is also at the same temperature.

2-3.   (a) Referring to Fig. 2-1, one might consider Types E, K, and T thermocouples, gas- or liquid-filled systems, or platinum resistance bulbs.

      (b) One might consider Types B, K, R, and S thermocouples, tungsten-rhenium thermocouples, or a radiation thermometer.

UNIT 3

3-1.   The Seebeck effect, which develops an emf if the hot and cold junctions are at different temperatures, and the Thompson effect, which allows the thermocouple to have varying temperatures along its length without affecting the output, if the thermocouple wires are of the same composition and state of annealing everywhere along their length.

3-2.   The emf output depends on the difference in temperature between the hot and cold junctions. The initial calibration of the measuring device assumes a temperature-emf relationship of the thermocouple based on a fixed cold junction temperature. Any changes in cold junction temperature, therefore, must be eliminated, or the measuring instrument must have a compensating means to make it read the same temperature as the hot junction temperature even though the cold junction temperature changes.

3-3.   Following from the Thompson effect, the law of the homogeneous circuit states that application of heat to a homogeneous wire will not cause an electric current, even though the cross section and shape of the wire may be different at different points along the wire.

3-4.   If a wire of a third metal is introduced in a thermocouple circuit, there will be no effect on circuit emf if the wire is homogeneous and the ends of the wire are at the same temperature. This law allows the use of copper wires between the thermocouple cold junction and the measuring instrument. The cold junction may be at a constant temperature bath or at the input terminals of the measuring instrument. Wires of other than thermocouple material may be introduced between the cold junction and the actual emf-measuring circuitry.

3-5.   The law of intermediate temperature states that the thermocouple emf caused by the difference between temperatures T_3 and T_1 is the sum of the emfs caused by the difference between T_3 and a temperature T_2, and by the difference between T_2 and T_1.

3-6.   Extension and compensating wires are usually used when they are less costly than running thermocouple wire.

Extension wires are less costly if they can be smaller than the thermocouple wire and if the lower temperature exposure of the extension wire permits a cheaper insulation to be used.

Compensating wire for noble metal (platinum) thermocouples is always less costly than thermocouple wire. Compensating wires for tungsten-rhenium thermocouples are less costly and, in addition, do not suffer from brittleness and oxidation as do the thermocouple wires.

3-7.   From Table 3-1, referenced to 32°F, the emf of the Type K thermocouple at 67°F is 0.776 mv.

The total thermocouple emf referenced to a cold junction temperature of 32°F is 0.776 + 8.054 = 8.83 mv.

Referring to the table, 8.83 mv lies between 423° and 424°F

	424°F	8.849 mv	8.830
	423°F	8.826 mv	− 8.826
$\Delta =$	1°F	$\Delta =$ .023 mv	.004

$$T_{HJ} = 423 + \frac{1(0.004)}{0.023} = 423 + 0.17 = 423.17°F$$

3-8.   Because the extension wire injects into the circuit an emf proportional to the cold junction temperature relative to the ice bath, the calibrator should be adjusted to the value read from the Table 3-1 based on 32°F cold junction temperature.

E = 8.314 mv

3-9.   From Table 3-1, referenced to 32°F cold junction, the emf for 400°F is 8.314 mv.

The emf for T_{CJ} = 69°F is 0.821 v.

Adjust the calibrator to 8.314 – 0.821 = 7.493 mv.

3-10. At 500°F the emf change per each °F change in temperature is approximately 0.023 mv:

498°F   10.514
499°F   10.537 Δ = 0.023
500°F   10.560 Δ = 0.023
501°F   10.582 Δ = 0.022
502°F   10.605 Δ = 0.023

The reading of –0.01 mv means that the reference thermocouple has a lower emf than the thermocouple under test. The thermocouple under test, therefore, has a higher emf than the reference thermocouple when both are at the same temperature.

(1) $\dfrac{0.010}{0.023}$ = 0.43°F. Therefore, the thermocouple under test

reads 0.43° higher than the reference thermocouple at the test temperature of 500°F. But the reference thermocouple reads 1.3°F low at 500°F. Therefore, with respect to the standard calibration the thermocouple under test is 1.3 – 0.43 = 0.87° low. The thermocouple under test is closer to the nominal calibration curve than the reference thermocouple.

3-11. Check each wire of the thermocouple under test with dry ice or a flame. Start with the wire at ambient temperature along the entire length, and look for a large change in emf between the ends of the wire as the cooling or heating source is moved.

Alternatively, repeat the calibration test with both thermocouples inserted at greater and lesser depths in the calibrating furnace. If the error changes significantly as the depth of immersion changes, the thermocouple under test is probably inhomogeneous.

UNIT 4

4-1. Moisture will cause the iron wire to rust, changing its calibration. Sub-freezing applications are likely to have moisture condensing on the thermocouple.

4-2. Type T and Type E thermocouples might be used.

4-3. A reducing atmosphere will shift calibration.

4-4. At temperatures above 1800°F (1000°C), the platinum may be attacked by silicon from the insulators or tube.

4-5. Tungsten-rhenium thermocouples are useful primarily for high temperature measurements in vacuum, inert, or reducing atmospheres.

4-6. Most sheathed thermocouples are rolled or drawn because these processes are less costly than swaging.

4-7. Integral, exposed, and insulated junctions are available.

UNIT 5

5-1.   The Type J thermocouple is a better choice. 600°F is within the range of both types. Because the application is outdoors, the connection head will be tightly sealed against the elements. Inside the well will be little air flow, so the atmosphere will probably be reducing.

5-2.   By pressing the hot junction against the end of the well or protection tube, a relatively constant thermal resistance can be maintained. In the absence of spring loading, thermal expansion and contraction may cause contact to change with time. Speed of response will certainly change, and accuracy may be affected if the immersion depth is low.

5-3.   When a thermocouple measures the temperature of a hot gas flowing in a duct at lower temperature, the thermocouple tip radiates energy to the duct and is at a lower temperature than the gas. The shield, at nearly the same temperature of the gas, results in less radiated energy being lost from the thermocouple.

5-4.   This would be a difficult measurement. The thermocouple, whether of foil or wire must have low mass compared to the sheet. A wire-type thermocouple would probably have to have a diameter of only a few thousands of an inch. If the temperature is low enough, an adhesive-type foil thermocouple might work if the mass of the assembly does not disturb the temperature distribution in the sheet. This might be the case if the sheet is much larger than the foil assembly and is at a relatively uniform temperature throughout.

UNIT 6

6-1.   Though copper is inexpensive, its low specific resistance results in bulbs that are either too large or of too low resistance for general application.

6-2.   Nickel was used to wind bulbs with high 0°C resistance, 300 ohms or more, because the measuring equipment could not readily handle lower values of resistance. Because the temperature coefficient of resistance of commercial nickel is not uniform from batch to batch, nickel bulbs had to be padded with fixed resistors to match a standard curve. Nickel bulbs were typically much larger than the platinum bulbs available today. Also, there was no industry-wide standard calibration curve.

6-3.   Two leads can be used if either of two conditions are met.

(a) The total lead resistance is negligible compared to the resistance change of the bulb at the desired accuracy.

(b) The measuring instrument can be calibrated to read correctly with the lead resistance in the circuit, and changes in temperature of the leads are small so the lead resistance remains constant, i.e., the change in lead resistance is small compared to the minimum change in bulb resistance at the desired accuracy of measurement.

Examples:

(a) From Table 6-1 at 200°C the resistance change of a 100-ohm platinum bulb is $\dfrac{178.51 - 175.84}{10}$ = 0.367 ohm/°C. If the measurement is not to be affected more than 0.1°C, the lead resistance must be less than 0.037 ohm. For 18-gauge copper, this is approximately 5.7 feet, so the bulb must be within 2.8 feet of the measuring instrument.

(b) The resistance of each copper lead of a 2-wire bulb is 5 ohms. At 0°C the total resistance is 110 ohms, and at 500°C the total resistance is 290.9 ohms. If the instrument can be calibrated to read 0°C at 110 ohms and 500°C at 290.9 ohms, the measurement will be accurate if the lead resistance doesn't change because of temperature changes of the leads. The permissible change in resistance for 1.0°C error at 200°C is 0.367 ohm. The 10-ohm copper lead resistance will change about 0.043 ohm for each degree change in temperature, so the temperature of the leads must be constant within ±8.5°C if a measurement accuracy of 1.0°C is to be achieved.

6-4. Tip-sensitive bulbs are spring loaded to increase speed of response. Other bulbs may be spring loaded to hold the bulb in contact with the protecting tube to reduce the likelihood of damage caused by vibration.

6-5. Assuming a bulb resistance of 100 ohms, the power dissipated in the bulb is $100 \times (0.0005)^2$ = 0.000025 w = 0.025 mw. The error is, therefore, less than 0.1×0.025°C = 0.0025°C, which is negligible in industrial practice.

6-6. Leakage resistance will lower the apparent resistance of the bulb at all temperatures, so the indicated temperature will be lower than the actual temperature. In some cases the error will increase with time.

6-7. If the fluid is well stirred so that the temperature gradients in the fluid are known to be small, reduce the depth of immersion of the bulb in successive steps. If the reading changes less than the desired accuracy of measurement, the depth of immersion is sufficient.

UNIT 7

7-1. As defined by SAMA:

    Class I — liquid-filled other than mercury
    Class V — mercury-filled
    Class II — vapor-filled
    Class III — gas-filled

7-2. Case-compensated systems utilize a bimetallic element in the instrument case to compensate for changes in case temperature. It is assumed that the capillary and case are at the same temperature.

Fully compensated systems use an auxiliary filled system without a bulb to provide a compensating motion proportional to temperature along the auxiliary capillary and in the case. Alternatively, a capillary system is designed with an internal steel wire so that temperature changes produce no net change in fill volume.

7-3.   Elevation differences must always be considered if the capillary is
        filled with liquid, as it is in liquid-filled systems and in Class IIA,
        Class IIC, and Class IID systems. However, in Class V systems, which
        are mercury-filled, elevation differences up to 30 feet generally do not
        produce significant error.

7-4.   Filled systems, compared to electronic systems, are limited in distance
        between the sensor and the instrument. Failure of a filled system bulb
        or damage to the capillary necessitates replacing the entire system.

UNIT 8

8-1.   Radiation thermometry is most likely to be used when the target is
        moving or cannot be touched, or when a high response speed is
        necessary.

8-2.   Compared to thermocouples and resistance thermometers, radiation
        thermometers are usually more expensive to purchase, install, and
        maintain. The selection of the proper radiation thermometer is more
        complicated than the selection of a resistance thermometer or ther-
        mocouple. There are no industry standard calibration curves.

8-3.   In conventional usage a total radiation thermometer is a broadband
        pyrometer that receives a high percentage of all the radiation emitted
        by the target.

8-4.   A ratio thermometer measures the energy in two narrow bands of the
        spectrum and computes the ratio of the two energies. This ratio is
        temperature-dependent. Any factor such as emittance, absorption, or
        target size, which affects both bands by the same percentage, will not
        affect the temperature reading. Ratio thermometers are useful for
        targets that do not always fill the field of view and installations where
        absorption due to water vapor or CO_2 varies, or where the emittance
        of the target changes by the same amount in both bands.

8-5.   In an optical pyrometer the radiation from the target at 0.65 μm, the
        red portion of the spectrum, is matched by matching its color to the
        radiation from a reference tungsten lamp. The match is made by the
        operator's eye. Either the temperature of the lamp is varied or the
        amount of target radiation is varied to make the two colors the same.
        When a color match is achieved, the temperature can be read from
        the adjustment dial.

8-6.   A blackbody source is an ideal radiator. It radiates the maximum
        amount of radiation of any body at a specific temperature. A black-
        body absorbs all the energy impinging on it. It has zero reflectance.
        Blackbodies are constructed by providing a chamber whose walls are
        at a uniform temperature and whose geometry provides multiple
        reflections of any ray entering it or being emitted from the wall.

8-7.   Any body emits energy to a receiver at absolute zero, according to the
        Stefan-Boltzmann Law, at a rate proportional to the fourth power of
        temperature. In practice the receiver is not at absolute zero, so the net
        radiation varies at a higher power of temperature depending on the
        range of wavelengths considered.

8-8.   Wien's Law is an approximation to the Wien-Planck Law where $e^{C_2/\lambda T}$
        is much greater than 1. It is easier to calculate the value of Wien's
        Law, and it also leads to the concept of the N value of a thermometer.

8-9.   The Wien Displacement Law says that the wavelength of maximum radiation becomes shorter as temperature increases. Thus, in general, at higher temperatures, thermometers with shorter wavelength response may be used.

8-10.   Emittance is the ratio of the energy per unit area that a real body emits in a specified band of wavelengths compared to the energy per unit area that a blackbody radiates in the same band of wavelengths, when both are at the same temperature.

8-11.   They are equal.

8-12.   As temperature increases, a material emits more strongly at the shorter wavelengths. Because metals usually have higher emissivity at shorter wavelengths, their total emittance increases with temperature.

8-13.   A thermometer measures over a band of wavelengths. It is possible to determine a single wavelength that would give the same change of output as the thermometer between two selected temperatures. This is the effective wavelength. The N factor or a thermometer can be calculated from the effective wavelength.

8-14.   A high N factor is always desirable because the approximate effect of changes in emittance is seen as a change in temperature indication by the emittance change divided by N. Therefore, an emittance change of 10% produces a temperature indication change of 10%/N (in absolute temperature).

8-15.   Choose an effective wavelength as short as possible. $N = \dfrac{C_2}{\lambda_e T}$, so the shorter the effective wavelength, the higher will be the value of N.

8-16.   The most common thermal detectors are thermopiles and bolometers. They are simple, reliable, and, when blackened, respond over a broad region of the spectrum. They are usually relatively slow to respond, and their output is low compared to photon detectors.

8-17.   Photon detectors respond to the energy of arriving photons and do not have to reach thermal equilibrium. They are, therefore, much faster in response than thermal detectors.

8-18.   The lenses or windows used in thermometers restrict the energy received to the region in which they transmit efficiently. Glass has a relatively low cutoff wavelength, so other materials must be used with detectors that respond at wavelengths longer than 2.5 μm.

8-19.   Radiation thermometers, in general, use large lenses to collect as much energy as possible. Radiation thermometers often work over a wide range of wavelengths. The larger the lens, the greater the spherical aberration. The broader the wavelength response, the greater the chromatic aberration. Both aberrations cause the focus of the thermometer to be somewhat fuzzy.

UNIT 9

9-1.   The speed of response needed, the target size and distance, and the target temperature will narrow the field of choice quickly in most situations.

9-2.   For most measurements an estimate of emittance from reference tables is sufficient. If the thermometer has a high N value, the error in indicated temperature will be acceptably small. Where higher accuracy is needed, a reference measurement with a thermocouple or optical pyrometer can be used.

9-3.   When temperature affects the emittance, by changing the state of a coating or thickness of an oxide film, or if the thermometer has broadband response so that the band emittance changes with temperature, the sample should be at normal target temperature.

9-4.   The energy lost by reflection from both surfaces should be calculated from the index of refraction of the window material at the wavelength of the measurement. For glass this loss is 8%. If the window has appreciable absorption, that loss too must be included.

9-5.   At the end of the heating cycle, the billet temperature and the furnace temperature are not far apart, so the reflected component of radiant energy is not greatly different from that which would be emitted at the true target temperature.

9-6.   A specular reflector reflects like a mirror. Float glass is a specular reflector. Some metallic sheets are also specularly reflecting.

9-7.   A radiation shield must be cooled if it can be heated by the surrounding radiation, which it is trying to eliminate from the measurement.

9-8.   Choose a thermometer whose spectral response lies outside the absorption bands of the gas or vapor. If practical, use a purged open-end sighting tube.

9-9.   If the particles are intermittently present, use peak or valley picking as applicable. If practical, use a purged open-end sighting tube.

9-10.   Unless the measurement can be made at a wavelength where the material is not transparent, the measurement is difficult because of the influence of hot objects or reflected sources behind the target.

9-11.   By comparing the unit under test to a reference standard of known accuracy and by making the comparison with both thermometers viewing the same source from the same distance, errors caused by fuzzy optics and absorption are the same for both thermometers. Also, the source need only be stable and uniform, not a high grade blackbody.

INDEX

Aberrations 121
Absorbency 103, 107

Band-pass thermometer 99
Bimetallic thermometer 141
Blackbody radiation 102, 106
Broadband thermometer 98

Calibration
    instrument 29
    radiation thermometer 136
    standards 19
    thermocouple 31
Celsius scale 13
Ceramic indicators 142
Classification of filled systems 87
Cold junction 22
    compensation 23
Color indicators 143
Compensationg wire 25
Copper resistance thermometers 73

Detectors 117
Direct temperature measurement 12
Disposable tip thermocouple 58

Emissivity 129
Emittance 107, 129
Expendable thermocouple 58
Extension assembly 52
Extension wire 25, 39
    insulation 41

Fahrenheit scale 13
Field of view 119
Filled systems 85
    application 93
    speed of response 93

Gas-filled systems 92
Glass stem thermometers 141

Heat balance thermometer 146
Hot junction fabrication 58

Immersion error
    resistance thermometers 77
    thermocouples 62
Inhomogeneity 33
Installing a thermocouple 61
Instrument calibration 29
Instruments 17
Insulation 41

Kelvin scale 13

Lenses and windows.. 120
Liquid-filled systems 87

Melting point indicators 143

Narrow band thermometers 99
N factor 110
Nickel resistance thermometers.. 72

Optical pyrometers 100
Optical systems 119

Peltier effect 27
Protecting tube 52, 54
Pyrometric indicators 142

Radiation pyrometer, see Radiation thermometer
Radiation thermometer
    aberrations 121
    accessories 136
    application of 129
    calibration.. 136
    construction 112
    field of view 119
    lenses and windows 120
    N factor 110
    selection 127
    signal conditioning 122
    speed of response 97, 118
    target size 128
    theory 110
    transparent targets 135
    types.. 98
Rankine scale 13
Ratio thermometer 100
Reflectance 107
Resistance thermometer
    accuracy 74
    copper 73
    immersion error.. 77
    insulation resistance.. 77
    lead configurations 71
    lead length errors 78
    measurement of resistance 72
    nickel... 72
    self-heating... 75
    speed of response 75
    standard calibration 68
    thin-film 72
    vibration resistance... 76
    wire-wound 67

Seebeck voltage 21
Selection
  radiation thermometer 127
  thermocouple 60
Self-heating... 75
Semiconductor sensors 145
Sheathed thermocouples 45
Signal conditioning 122
Speed of response
  filled systems 93
  resistance thermometers 75
  radiation thermometers 97, 118
Stefan-Boltzmann Law 104
Surface temperature measurement 56, 116

Temperature measurement
  direct 12
  indirect 12
  objective 11
  surface 56, 116
  useful range of sensors 17
Temperature scales 13
  conversion between 14
Temperature standards 15
Terminal block.. 57
Thermistors 145
Thermocouple
  accuracy 39
  calculations 26
  cold junction 22
  disposable tip 58
  hot junctions 46
  inhomogeneity 33
  installation-caused errors 61
  insulation 41
  selection 60
  sheathed 45
  types... 39
Thermoelectricity 21
  laws of... 24
  Peltier effect 22
  Seebeck effect 22
  Thompson effect 22
Thompson effect 22

Vapor-filled systems
  construction 89
  errors 91

Wells 52, 55
Wien-Planck Law 105
Wien's Law 106